QUICK
&
EASY
STATISTICS

A Practical and Interactive Approach Using SPSS

FATAI AKEMOKWE

Disclaimer: The author has made the best effort to provide accurate information on the subject matter covered. However, the author and publisher assume no responsibility for any errors or omissions. By reading this book, you acknowledge that you assume full responsibility for the use of the materials and information contained herein. Under no circumstances will the author or publisher be held liable for any loss or damage caused by your use or reliance on the information published here or contained in resources referenced in this book. This includes any websites, software, books, eBooks, or any other materials associated with this publication.

For 'Misan,
for making every moment meaningful.

CONTENTS

ACKNOWLEDGEMENTS

My thanks go to Dr Afe Orilade and Jafaru Akemokwe, my co-instructors at Enigma Ventures. Dr (Mrs) V. Omuemu and Dr. Frank Imarhiagbe mentored me in clinical research; you are always appreciated.

This book only exists because of all the researchers who have pushed me into seeking simpler ways to explain Statistics using SPSS. Thanks for the pressure.

WELCOME

T he aim of this book is to provide you with a focused practical approach to the use of statistics in research with emphasis on use of the IBM® SPSS® Statistics as a tool. At the time of writing, the latest SPSS available is version 22. Differences in version so far have not altered the basic process of using it for analysis.

For many aspiring researchers, learning statistics is a frightful rite of passage. This book sticks to the essentials. I have trimmed away the distracting and often unnecessary aspects that are best left to career statisticians.

Although the emphasis here is on using SPSS, the necessary theoretical basis in statistics will be provided as succinctly as possible. This knowledge can then be adapted to other statistical software.

The sample data files used in this book can be downloaded using this Internet link: is.gd/zMDZ1D Unless otherwise stated, *myData.sav* is the data file used in the examples.

Visit our website and blog for these and more resources.

Enjoy.

AN INTRODUCTION TO STATISTICAL TERMS

There are 3 types of lies:
lies, damn lies and statistics!
- *Attributed to Mark Twain*
(or Benjamin Disraeli.
Apparently, history is the fourth type of lie.)

Statistics is the science (an, maybe, art) of collecting, organizing and analyzing data.

Descriptive statistics: aims at describing a population:

Inferential statistics: aims at drawing conclusions from the data.

Data: raw facts and figures.

Population: the entire set of data which one aims to describe or make inferences about.

Sample: A subset of the population

Variable: any piece of data that varies from case to case or varies from one observation to another. (A *constant* is data that remains unchanged)

Independent (or Predictor) variable: a factor whose effect on another variable is to be assessed. This is also called a **Risk Factor**, **Explanatory variable** or **Exposure variable**.

Dependent (or Outcome) variable: a variable whose value is assumed to be influenced by other variables.

For example, does the sex of patients affect their response to antihypertensive medication? In this case, *sex* is the predictor variable while *"response to antihypertensive medication"* is the dependent variable. On the other hand, if the research question is *"Does timing of intercourse affect the sex of the fetus?"*, then *"sex of the fetus"* becomes the outcome variable while *"timing of intercourse"* is the independent variable.

Parameter: a number that is used to describe the *population*. It is represented with Greek letters for example, population standard deviation (σ), population mean (μ).

Statistic (without the final *"s"*!): a number used to describe a *sample*. Represented with Roman (normal) letters for example, sample standard deviation (s), sample mean $(\bar{x})$.

STUDY DESIGN

The design of a study determines the analysis to be performed on data that is collected. In turn, choosing a study design depends on the aim of the researcher.

Do you intend to describe the characteristics of one or more populations in terms of rates, proportions and percentages?

> Study design: (Descriptive) Cross-sectional/Prevalence study
>
> Example of Research Question: *How many doctors work in Gwagalada? How many of them smoke tobacco? What is the prevalence rate of hepatitis B antigen positivity in nurses in Gwagalada?*

Do you wish to analyze the frequency of occurrence of a suspected risk factor in those with a disease ("cases") compared to its frequency in those without the disease ("controls")? *Controls may be randomly selected or they be matched (in terms of age, sex, race or other variables).*

> Study Design: (Retrospective) Case-Control study
>
> Example of Research Question: *When*

compared to a group of healthy persons, are lung cancer patients more likely to have smoked cigarettes?

Do you plan to follow up healthy persons over time to determine the risk factors associated with future occurrence of a disease?

Study Design: (Prospective) Cohort Study

Example of Research Question: *Over a thirty-year period, are smokers in Gwagalada more likely to develop lung cancer compared to non-smokers?*

Do you want to find out how effective an intervention is?

Study Design: (Experimental) Randomized Control Trial

Example of Research Question: *Is there any difference in the efficacy of radiotherapy, chemotherapy, radical surgery, or various combinations of these therapies in the treatment of early-stage breast cancer?*

PREPARING YOUR DATA FOR ANALYSIS

TYPES OF DATA/VARIABLES

All data boils down to two *types*:

Numeric: any data that is represented with numbers only. Formats available in SPSS are:

(Plain) *Numeric*

Date: Allows dates to be entered in various formats

Dot: Decimal points represented with dots, every thousand separated by commas (as used in Nigeria) for example, 999,999,999.99

Comma: Decimal points represented with commas, every thousand separated by dots (as used in France) for example, 999.999.999,99

Scientific: for example, 3.2E3 representing 3.2×10^3 (3200) while 3.2E-3 represents 3.2×10^{-3} (0.0032)

Dollar: used to represent currency in US $

Custom *Currency*: Other currency

String: any combination of letters, symbols or numbers String and ordinal data make up **categorical** data. Data may also be "**Missing**" where the variable does not apply to the subject (for example, prostate size in a female) or, for some reason, was not obtained *ab initio*. "Missing" data is excluded from most analyses.

Measures of variables in SPSS may be:

Nominal: data belonging to categories that are mutually exclusive for example, sex, marital status, occupation. They are usually represented by text (**String**) but may be represented by numbers for ease of analysis. However, these numbers have no quantitative importance (for example, male=1, female=2 or male=0, female=1). Data with only two (2) categories (examples: "Yes/No" questions, Sex) is called *binary* data. In SPSS, nominal variables are preceded by the icon

Scale: Measurable data with fractions for example, weight, height, dose of medications. These values can be added, subtracted, divided or multiplied. They may also be negative or positive in value. In SPSS, scale variables are preceded by the icon .

Ordinal: data that can be ranked for example, position in class, degree of heart block, stage of cancer. One rank is higher than the next but there is no way of dividing, multiplying, adding or subtracting (someone with Class II heart failure cannot be said to have twice as much heart failure as someone with Class I!). Ordinal data is represented by whole numbers in SPSS; it does not allow for decimal points or fractions. Ordinal variables in SPSS are preceded by the icon

CREATING A DATASET

There are two options in creating a dataset (also called a database) in SPSS. You may choose to enter the data directly into SPSS or import the data from another source.

ENTERING DATA DIRECTLY INTO SPSS

Prior to entering data directly into SPSS, ensure that the following characteristics of each variable have been written out in a **code dictionary** based on your research tool/questionnaire.

Name	Label	Type (Width*)	Codes for Values	Codes for Missing
Serial_No	Serial Number	Numeric (3)	None	None permitted
Sex	Gender	Numeric (1)	0 = Male 1 = Female	9
Age	Age (Years)	Numeric (3)	None	999
MarStat	Marital Status	Numeric (1)	0 = Married 1 = Never Married 2 = Divorced 3 = Widowed	9

An example of a coding dictionary.

*Width=number of characters SPSS will accept in the cells (a width of "3" will truncate a value of 4567 to 456, causing confusion!)

RULES FOR CODING

---

1. Variable names must start with a letter. They should NOT contain spaces or special characters (like (, *, &, %, $, #, @).

2. Each variable name must be unique (no duplication).

3. The Variable Label is what appears in the tables, graphs and other output after analysis. It can contain spaces and any other character. for example, "Age of Respondents (Months)"

4. As much as possible, use numbers to encode data. This allows for ease of analysis using SPSS.

5. When coding, the positive response in a predictor variable should have the higher numerical value for example, Smoker = 1, Non-smoker = 0 (or Smoker=2, Non-smoker = 1). If you are looking at male sex as risk factor, then code male = 1, Female =0. On the other hand, if female sex appears to be predictive, then code female=1 and male=0.

6. Similarly, when coding for outcome variables, the outcome we are interested in should have the higher numerical value for example, does excess television viewing increase the risk of in-patient death at the 30th day of admission? Here, the

outcome variable name could be **Death_30**, with "**Dead**" coded as 1 and "**Alive**" coded as 0 since we are primarily concerned with those that died.

7. SPSS is case-sensitive. If you are going to use text (for "**String**" variables), it is best to ensure that the data values are entirely in small-letters or capital letters. "Cat", "cAT", "CAT", "cat" and "CAt" would be recognized as three different values by SPSS.

8. The essence of coding is to simplify data entry. Avoid long codes.

9. Select an impossible value as the code for missing values for example, negative or extremely large values for "Age".

10. Where data contains unmatched/unpaired groups you want to compare (cases vs. controls), a special *grouping* variable can be created (for example, variable name: Group1, Cases coded as 1, Controls as 0). This allows for easy comparison of such groups using hypothesis testing.

11. For paired/matched data, different columns should be created for each variable. An example is where blood pressure is measured for subjects and subsequently the measurement is repeated. The readings can occupy two columns as distinct variable (probably named BP1 and BP2)

THE SPSS INTERFACE

On running the SPSS program, typically a dialog window appears.

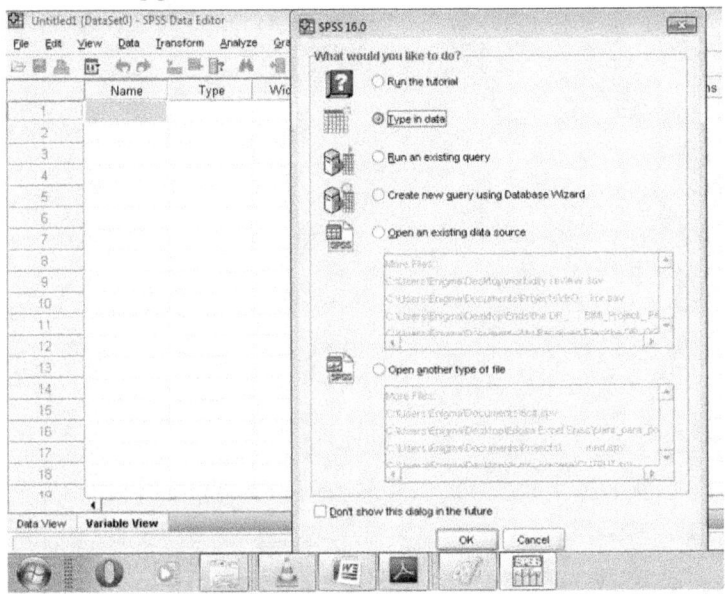

If you choose the "Type in data" option, a Data Editor interface follows.

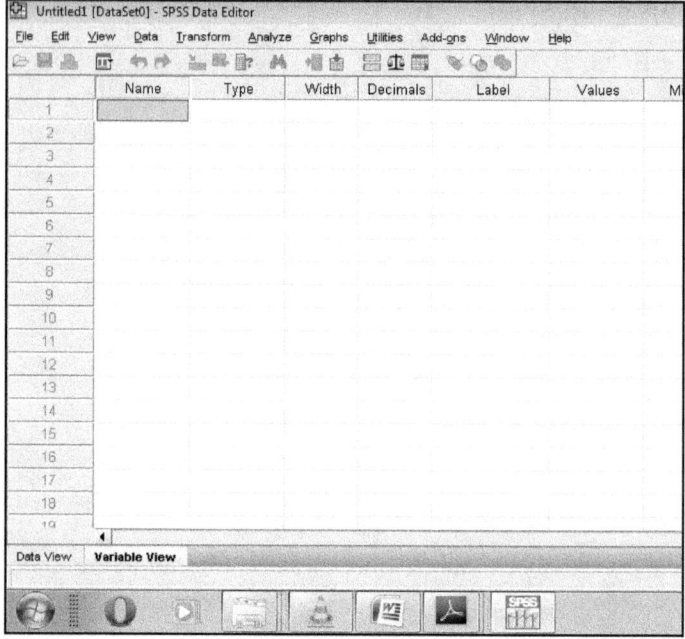

The **SPSS Data Editor** interface has two windows: a **Data View** and a **Variable View**. The next step is to transcribe the variables defined in the coding dictionary into the **Variable View**.

• Click on the tab marked "**Variable View**". Enter the variable names, label, type, decimal (number of decimals, default value is 2) and width (default value is 8).

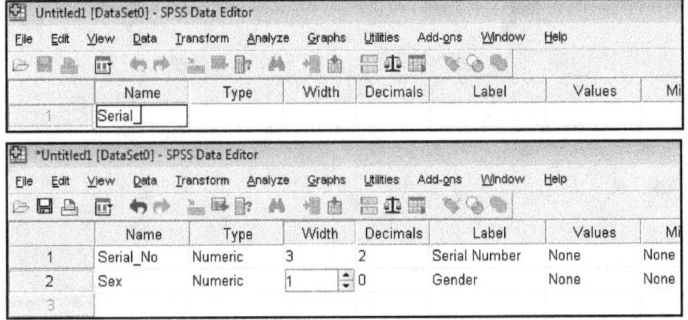

- ## Define the values for each variable.

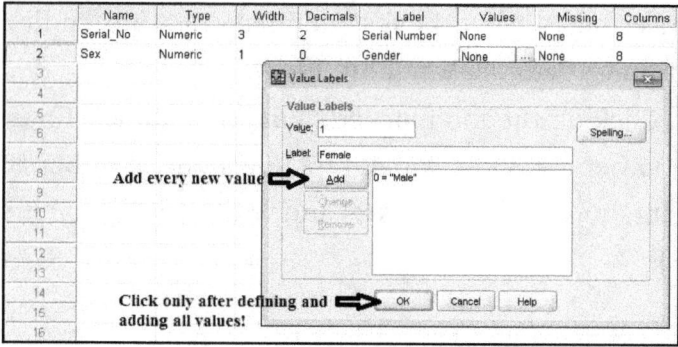

	Name	Type	Width	Decimals	Label	Values	Missing	Columns
1	Serial_No	Numeric	3	2	Serial Number	None	None	8
2	Sex	Numeric	1	0	Gender	None	None	8

- Define missing values.

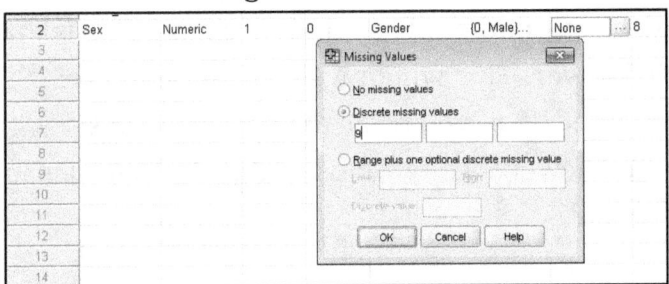

After defining your variables, save the data set. Do this frequently to prevent data loss. Go to "**File**" >>"**Save**", or simply use the 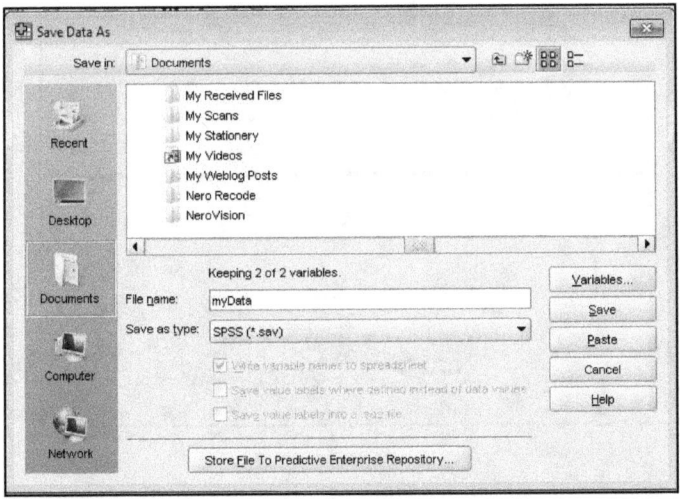 icon.

17

Datasets are saved in the *.sav* format. After saving the dataset, another SPSS window opens. This is the **SPSS Viewer**. This window shows a log of all SPSS actions and it is where all results of analysis will appear. Viewer logs can be saved as Output files in *.spv* format.

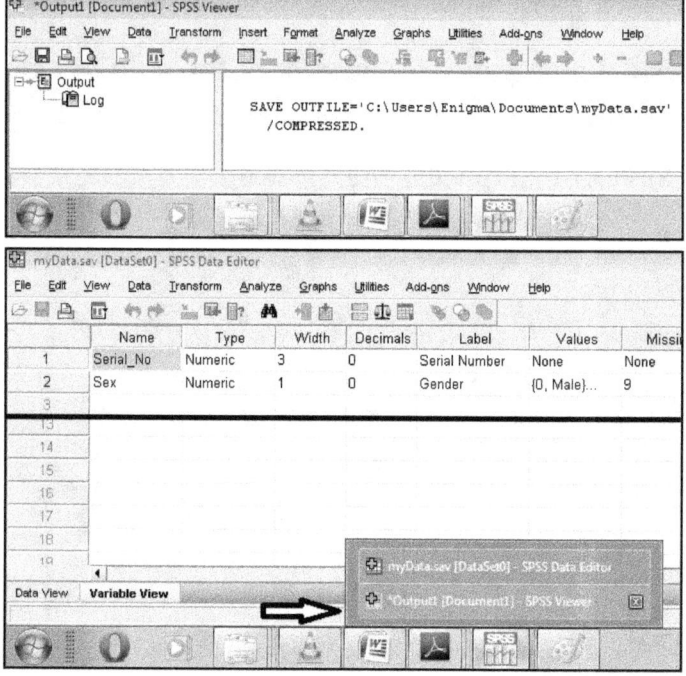

Return to the Data Editor Window and click the "**Data View**" tab at the bottom-left corner. The Data View is a spreadsheet with vertical columns ("**Variable**") and horizontal rows ("**Cases**") which intersect at cells. This is similar to the spreadsheet in Microsoft Excel® but there are a few important differences. Unlike Excel, cells can only contain values, not formulas. Hence, values within cells do not update automatically when changes are made to related cells. Unlike Excel where variable names are entered in the topmost row, in SPSS, this contains the data of the first subject. You can then type in the data into the appropriate cells using the codes.

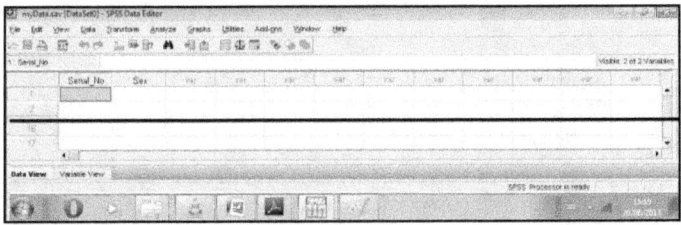

ELEMENTS OF THE SPSS INTERFACE

The visual elements we come across when using SPSS include:

1. Main Menu

File   Edit   View   Data   Transform   Analyze   Direct Marketing   Graphs   Utilities   Add-ons   Window   Help

2. Drop Down Menus

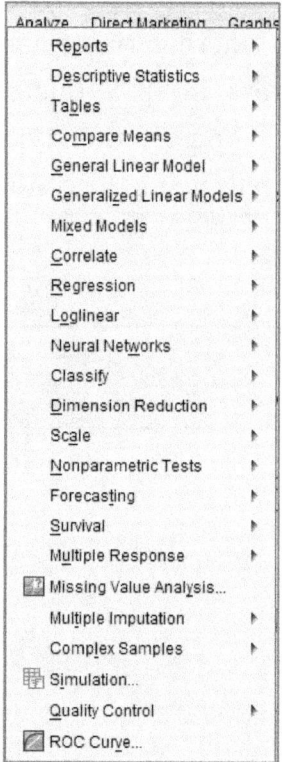

3. Dialog Window

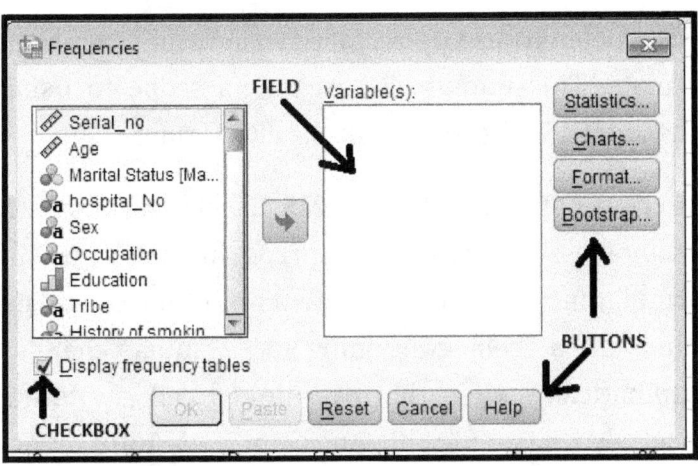

4. Radio Buttons

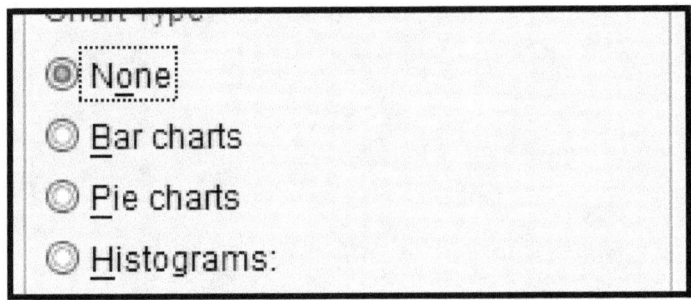

IMPORTING DATA FROM OTHER SOURCES

SPSS is able to use data stored in several other data formats. This book will restrict its scope to using data stored in the Microsoft Excel® format.

Entering data initially into MS Excel® has several advantages. The cells in the spreadsheet allow use of formula linking them to contents of other cells. The value of a cell containing a formula changes automatically when that of a linked cell is altered. Excel also has AutoFill, AutoSum and AutoCorrect functions which can be time-saving during data entry.

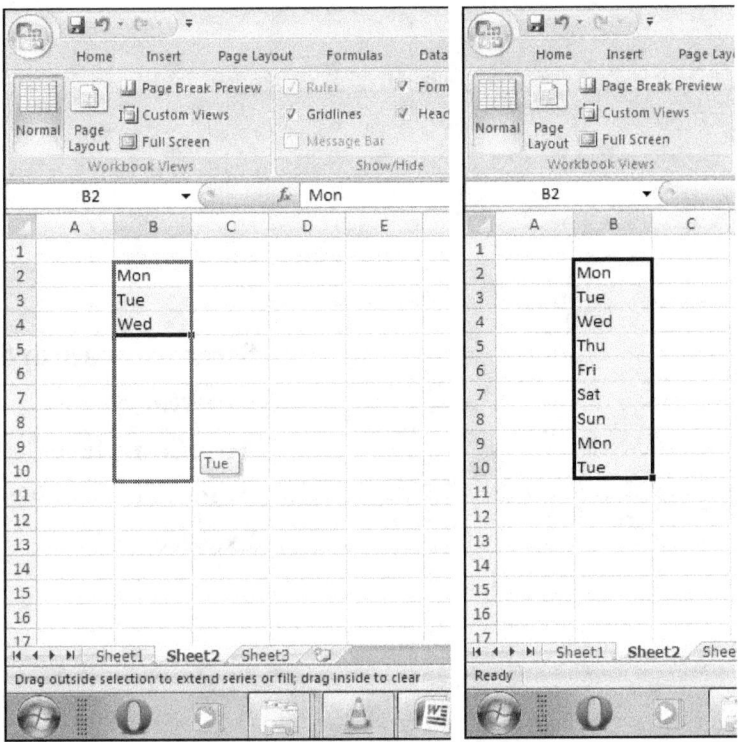

AutoFill in MS Excel®

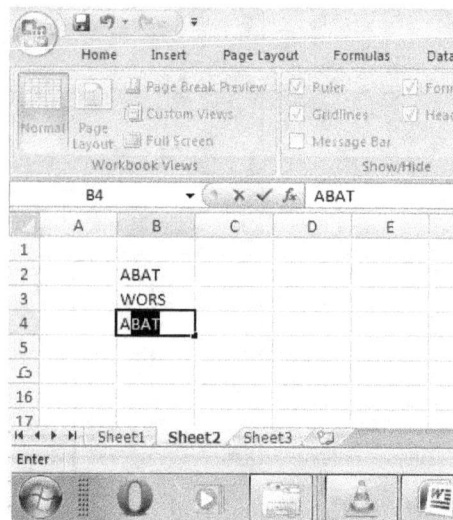

Screen capture showing AutoComplete function in MS Excel®

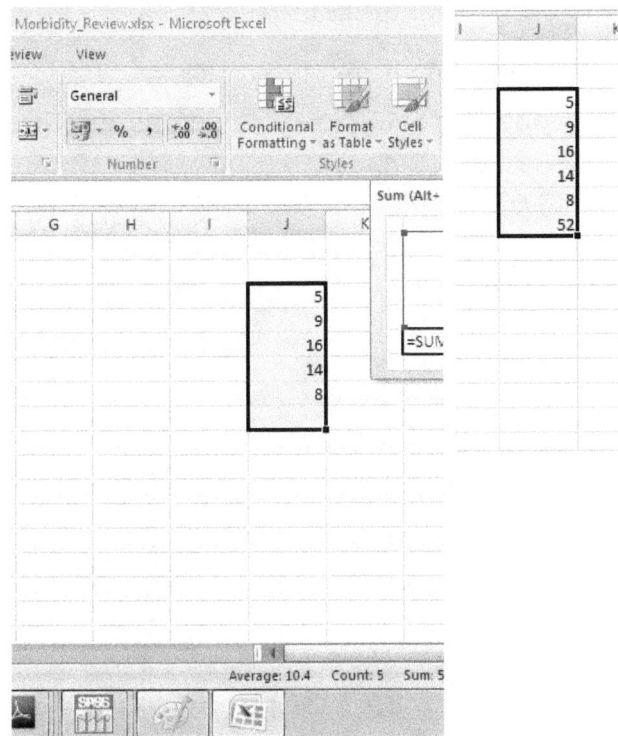

AutoSum function in MS Excel®

To import existing data from MS Excel®:

1. On first opening SPSS Select "Open an existing data source" >> "More Files…"

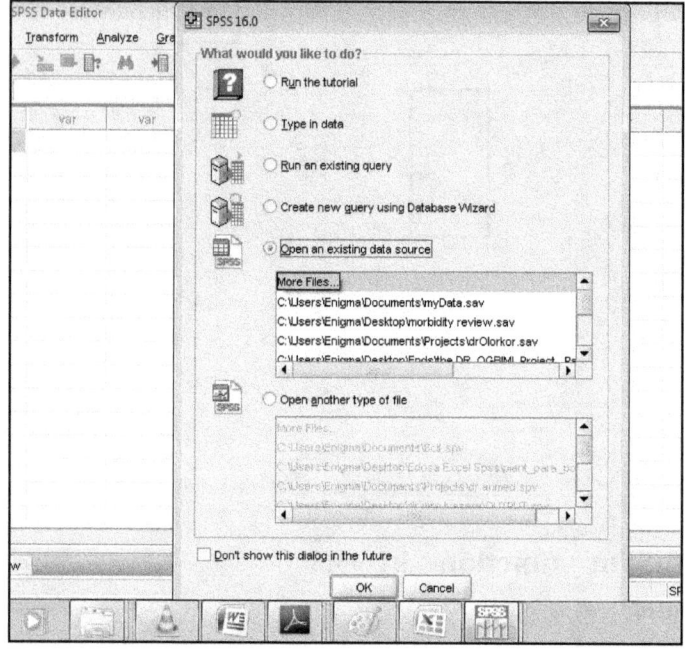

2.   When the dialog window opens, navigate to the folder containing the desired Excel file. Click the drop-down menu labelled "**Files of type:**" and select "**Excel (*.xls, *.xlsx, *.xlsm)**". Double-click the desired file.

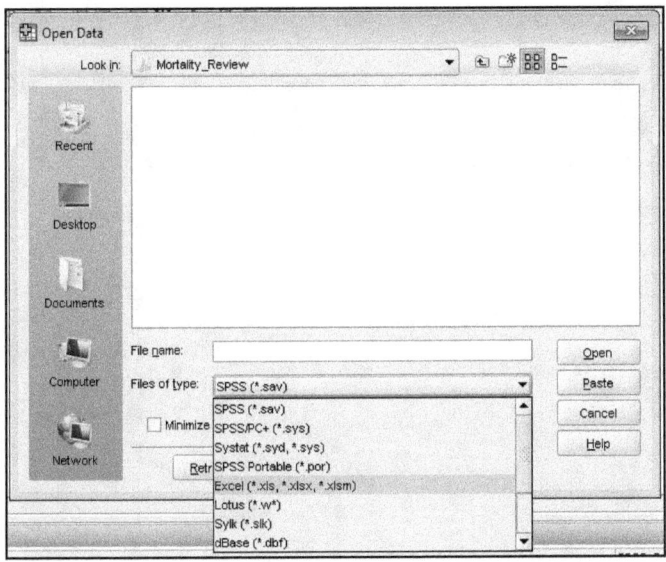

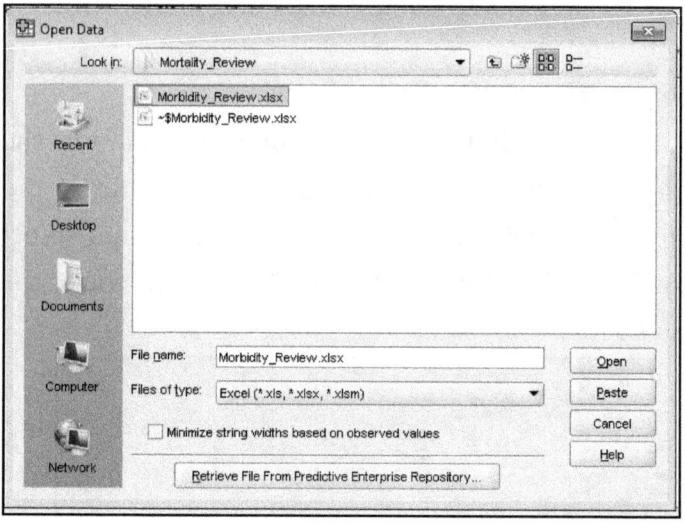

3. Another dialog box appears. Ensure you have ticked the checkbox on "**Read variable names from the first row of data**", then click "**Continue**"

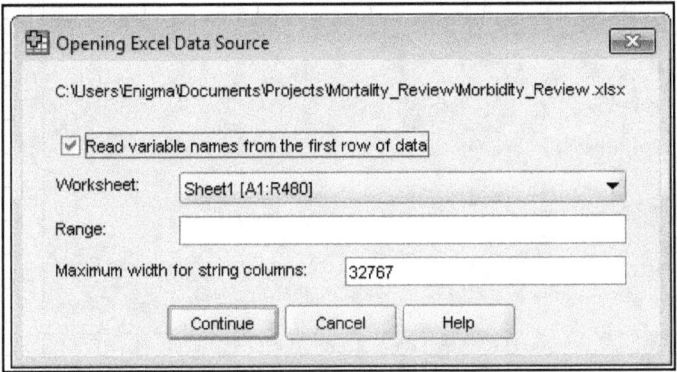

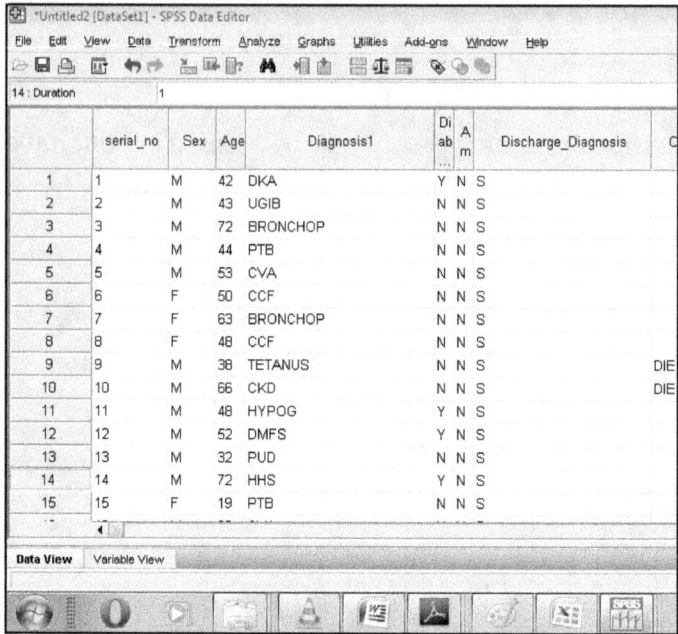

4. Save the dataset as earlier discussed. Note that if the variable name in Excel is invalid, it will be replaced with "**VAR00000XX**" where X is a number. If the variable name in Excel is a special SPSS command (for example, GET, FREQUENCY), the variable data will not be imported and an error message will be displayed in the viewer. Blank cells in Excel become "**System-Missing**" values in SPSS, displayed as full-stops.

5.   Use Variable View to define the label, variable type, values codes and missing values code.

*An alternative method is to go to **"Files"** >> **"Open"** >> **"Data"**. Then, go through steps 2 to 5 listed above.*

OTHER TASKS IN DATA HANDLING IN SPSS

ALTERING DATA ENTRIES

To change a data entry, select the cell, type in the new value and press the **Enter** key. This overwrites the previous content of the cell.

To delete the content of a cell, select the cell and press the **Delete** button on your keyboard. Entire cases or variables can be deleted by selecting the grey headers and pressing the **Delete** key. Alternatively, right-click the header and select "**Clear**".

Any changes can be reversed using "**Undo**" icon or holding **CTRL+Z.** Changes do not become permanent until the file is saved.

COPYING AND PASTING DATA

Data in single rows, single columns or across both can be copied or cut from one part of the worksheet and pasted in other parts of the spreadsheet. Tabular data from other programs (including 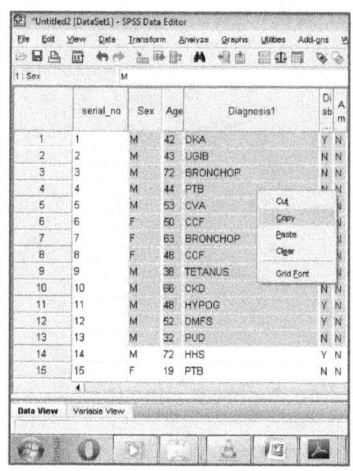 Excel) can also be pasted into SPSS. Note that pasted data will overwrite the contents of the destination cells.

To copy or cut data,

1.   Select source cells

2.   Hold **CTRL+C** (to copy) or **CTRL+X** (to cut). Alternatively right-click and choose "**Copy**" or "**Cut**". A third option is to go to "**Edit**" > "**Copy**"/ "**Cut**"

Select the cell at the upper left corner of target destination and hold **CTRL+V**. alternatively right-click and choose "**Paste**". A third option is to go to "**Edit**" > "**Paste**"

INSERTING NEW CASES/ VARIABLES

To insert a new variable (i.e. a new column), select the column that will be on the right of the new variable. Right-click and select "**Insert Variable**". Alternatively go to "**Edit**" >> "**Insert Variable**". The third option is to click on the icon .

To insert a new case (i.e. row), select the row that will be below the new case. Right-click and select "**Insert Case**". Alternatively go to "**Edit**" >> "**Insert Case**". The third option is to click on the icon .

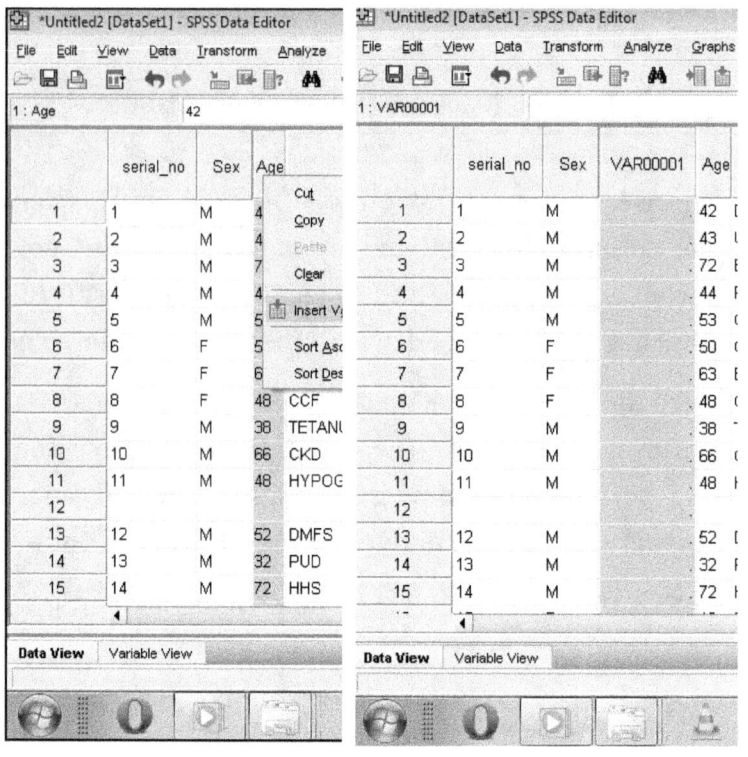

MOVING DATA

To move a variable, *left*-click the column header and *drag* to the target location using the red vertical line as the marker.

To move a case, left the row header and drag similarly using the horizontal red line as a marker.

3	3	M		3	3	M
4	4	M		4	5	M
5	5	M		5	6	F
6	6	F		6	7	F
7	7	F		7	8	F
8	8	F		8	9	M
9	9	M		9	4	M
				10	10	M

COMPUTING A NEW VARIABLE

The aim here is to create a new numeric variable from previously existing numerical variables. In this example, we shall create a new variable "**BMi**" (Label = Body Mass Index) from two previously defined variables called "**Weight**" and "**Height**" using the formula BMi = Weight/Height^2.

To do this, go to "**Transform**" >> "**Compute Variable...**" >> Type in Variable name in the text field "**Target Variable**"

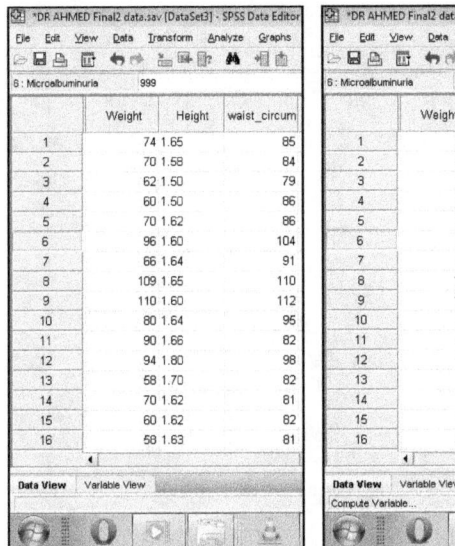

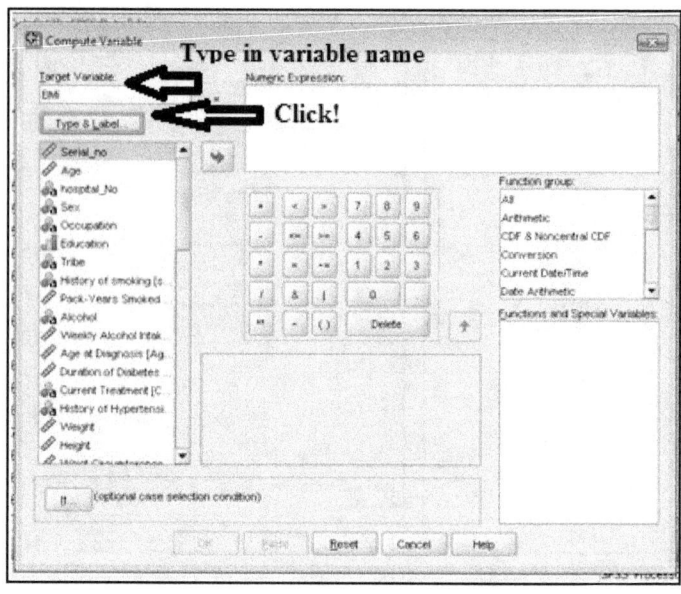

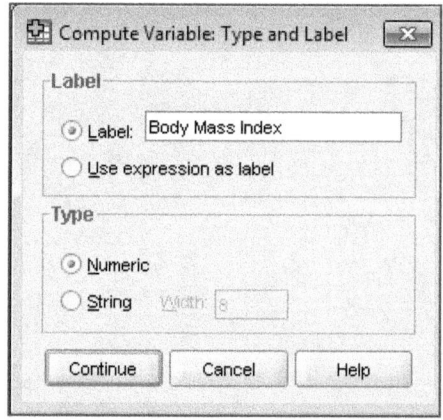

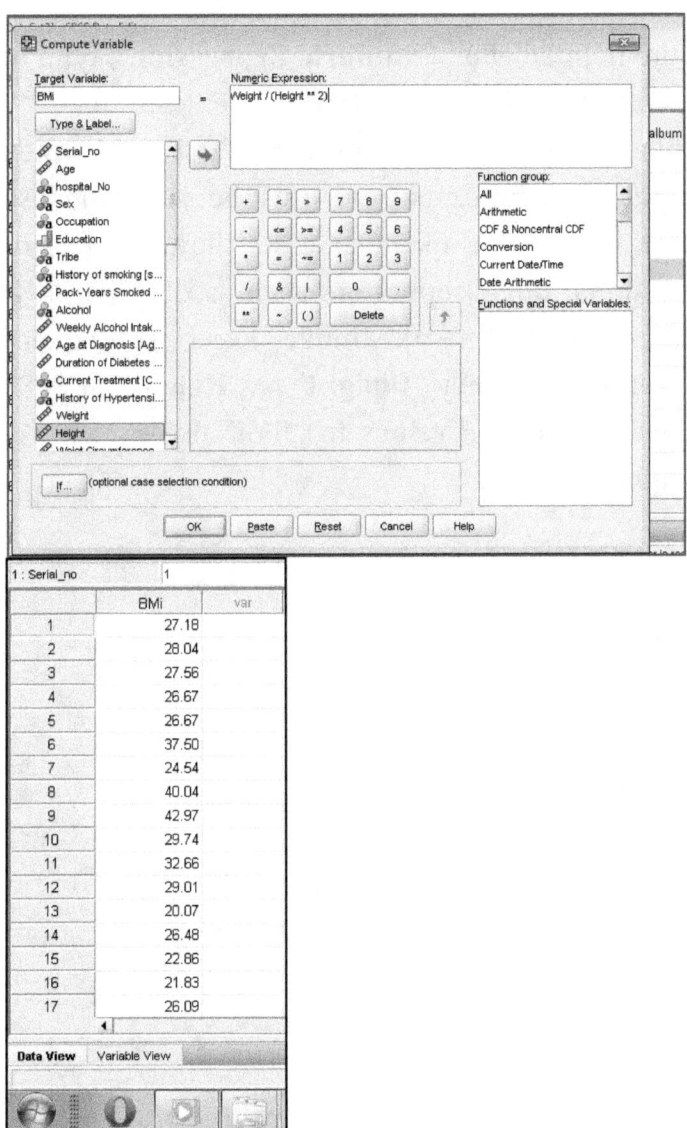

In the field labelled "**Numeric Expression**", type in the formula using the BODMAS rule. Note that "**" implies "raised to the power of". For complex formulas, use brackets freely (just be sure you close all of them). The formulas also allow for use of conditional (logic) operators such as OR ("|"), AND ("**&**") and NOT (~=). Unlike Excel, subsequent alterations in either "**Weight**" or "**Height**" in SPSS will not affect the values in "**BMi**" unless it is re-computed.

RECODING A VARIABLE

The purpose of this is to transform one variable into a new variable ("**Recode into Different Variable**") or to replace its values ("**Recode into same variable**") using certain criteria. Please note that "**Recoding into *same* variable**" causes the previous values to be discarded leading to possible loss of important data.

In this example, our newly generated variable "BMi" will be recoded into a different variable called "**DegObese**", for "**Degree of Obesity**". The values will be 1 (BMi <18.5), 2 (BMi between 18.5 and 25.9), 3 (BMi between 26 and 29.9) and 4 (BMI of 30 and above).

Go to "**Transform**" >> "**Recode into different variable**". The following dialog window appears. From to the list to your left, select the variable to be recoded, and transfer it to the field marked "**Input variable→Output Variable**". Type in the desired name and label of the output variable in the text fields under "**Output Variable**", click "**Change**" and "**Old and New Values…**"

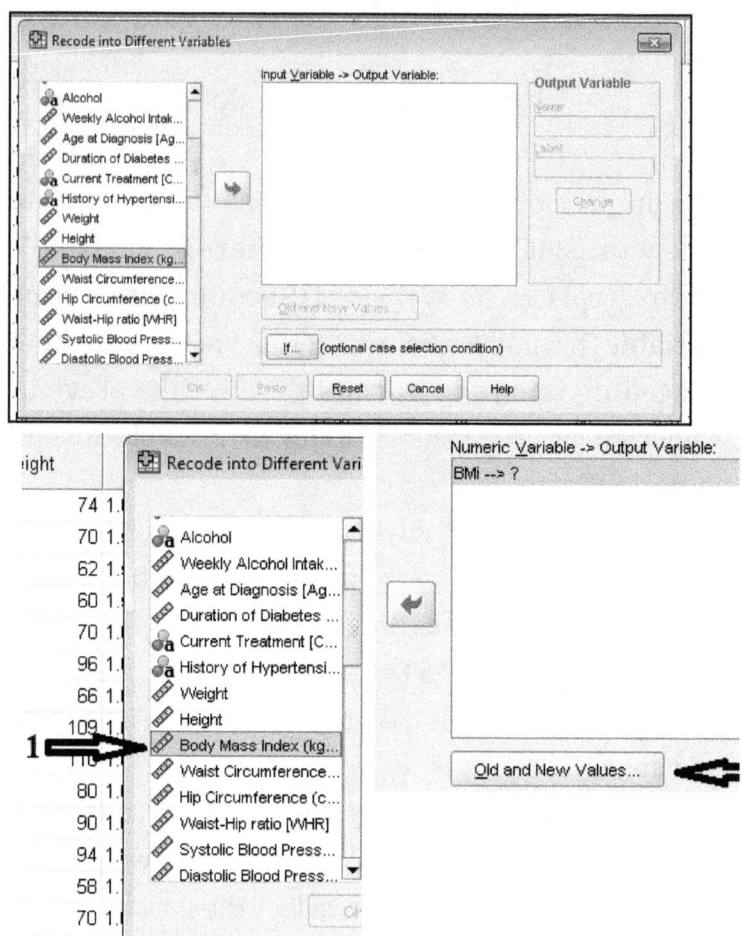

At this stage, another dialog window opens for the actual recoding. To recode **BMi** <18.5 into **DegObese**=1. (Note the use of *18.4* instead of 18.5. This is because typing 18.5 would cause subjects with BMI of 18.5 to be included and wrongly categorized as "1" instead of "2".)

The "**New Value**" field, by default, accepts only numbers. If the desired output variable is "**String**" (for example, **DegObese**= "**Underweight**" instead of 1), tick the first checkbox to allow typing of text.

If numbers were imported as **String** (probably unintentionally), then they can be automatically recoded to numeric data by ticking the second checkbox.

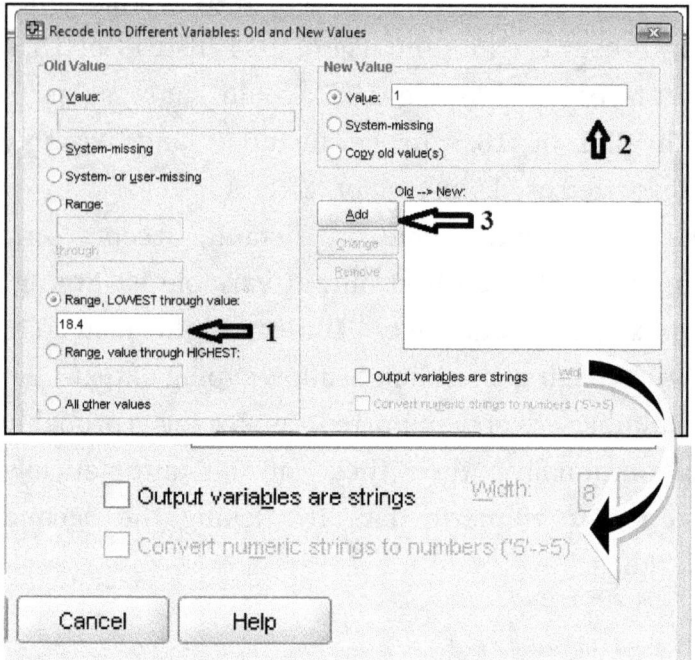

To recode **BMi** between 18.5 and 25.9 into **DegObese**=2: (Note the use of *18.5* here as this value is included in this category!)

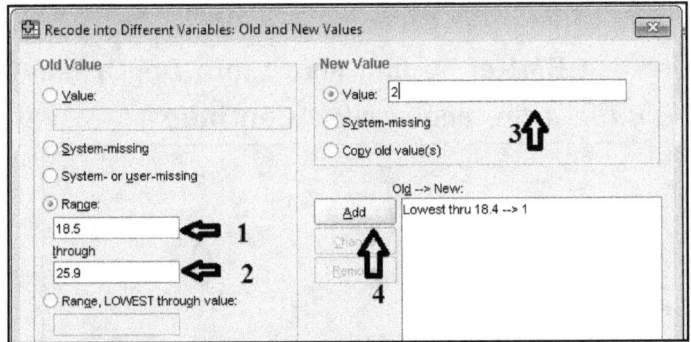

To recode **BMi** between 26 and 29.9 into
DegObese=3:

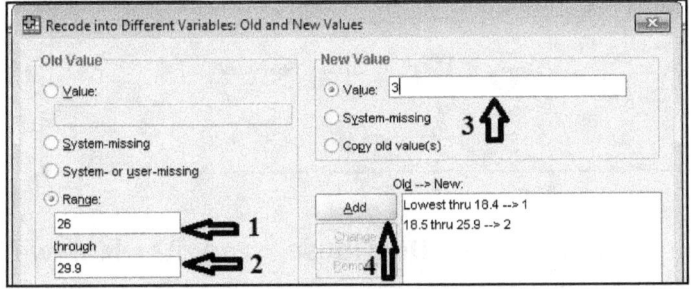

To recode **BMi** of 30 and above into **DegObese**=4
(*This is the last recode, so click* "**Continue**")

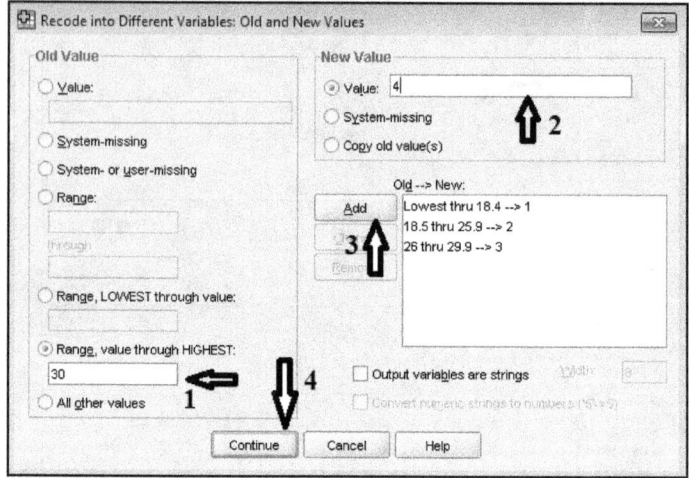

The new variable "**DegObese**" and its data will appear on the **Data Editor**.

Don't forget to save your work frequently!

HANDLING MULTIPLE-RESPONSE DATA

So far, we have dealt with data for which every variable has a single value for each case. What if the research tool/questionnaire allows selection of multiple options for a particular variable? An example is in surveys where a question like "Source of health information" may produce responses like **"Newspaper"**, **"Newspaper, Seminar"**, **"Seminar, Hospital, Newspaper"**.

To get around this, each response should be coded as a separate variable with values of 0 and 1. A value of 1 may be chosen to imply that the case selected that particular variable in which case a value of 0 implies that this was not one of the case's choices. (Note that the choice of values is arbitrary. The only condition is that they must be numbers.)

All the variables representing multiple responses to a single question are then defined in a Multiple Response Set. To do this, go to **"Analyze"** >> **"Multiple Response"** >>**"Define Variable Set..."** In the dialog window that appears, select the variables representing all the responses and type in the **"Counted Value"** (in this case, **"0"**).

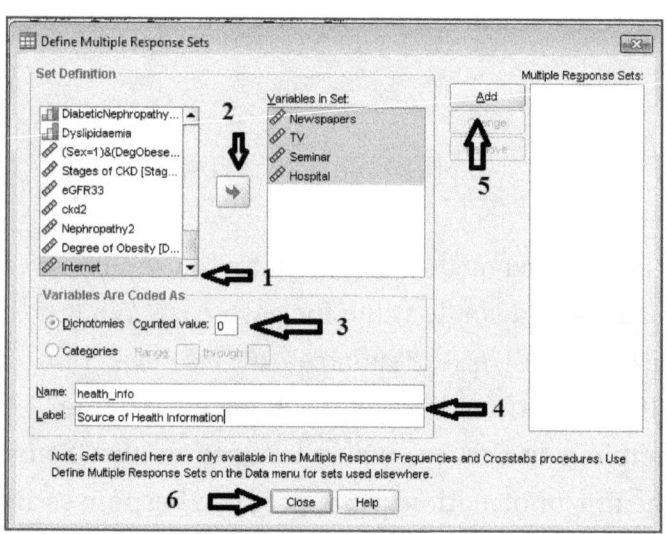

SELECTING OR FILTERING DATA

Assuming you want to run analysis on only the obese females in your sample, how do you get SPSS to ignore the males and non-obese females? You need to instruct SPSS to *select* cases only *if* gender ("**Sex**") equals female (1 in this case) AND if subject is obese (**DegObese**=4).

To do this, go to "**Data**">> "**Select Cases**".

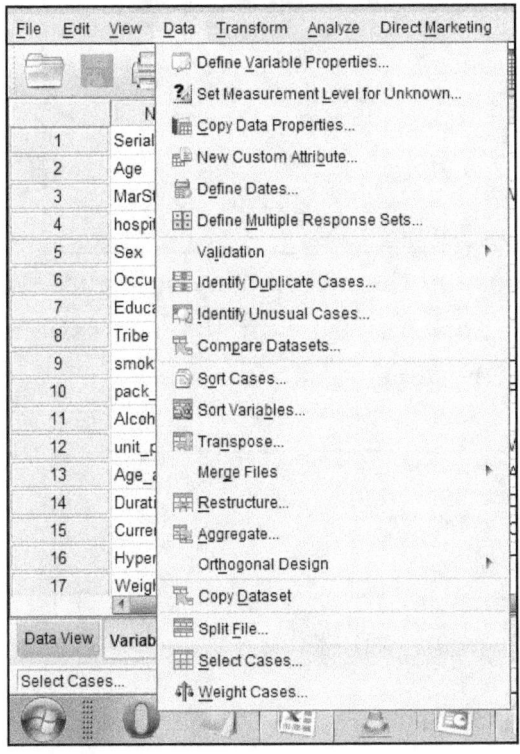

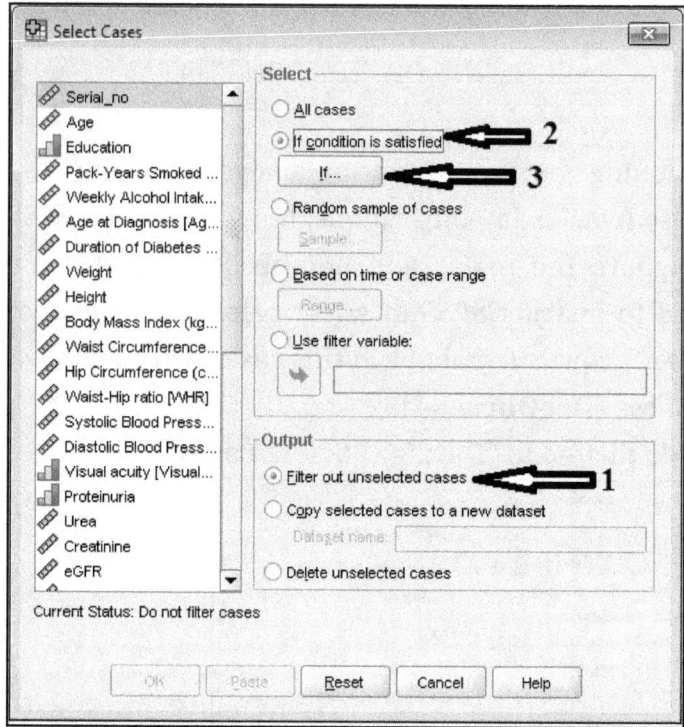

Follow the steps in the accompanying figure. A dialog window appears. Type in the expression: (**Sex**=1) & (**DegObese**=4).

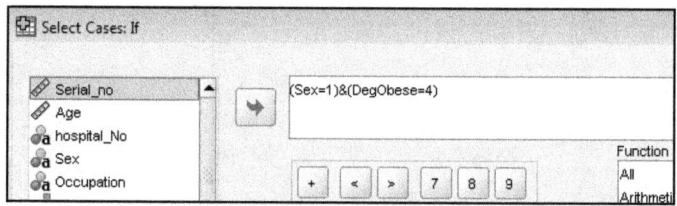

Click **"Continue"** >> "**Ok**".

Unselected (filtered) cases will have a line across their row headers.

	b	Age	hospital_No	Sex
1	1	58		1
2	2	45	159806	1
3	3	58	236204	1
4	4	61	226317	1
5	5	48	427112	0
6	6	56	536144	0
7	7	44		1
8	8	44	487214	1

To undo this filtration/selection, go to "**Data**">> "**Select cases...**" and select the radio-button marked "**All cases**".

DESCRIBING DATA

The aim of describing data is to reduce it to manageable chunks that still convey valid and relevant information.

EXPLORING YOUR DATA

A preliminary overview of numerical data obtained provides insight into errors (a subject's age of 200 instead of 20) and omissions. It also enables the researcher determine if certain variables are normally distributed or skewed. Overall data can be explored initially followed by exploration of subsets of data (grouped by sex, age group, etc).

The examples in this chapter use data from the file *myData2.sav.*

To explore data, go to "**Analyze**">> "**Descriptive Statistics...**" >> "**Explore**".

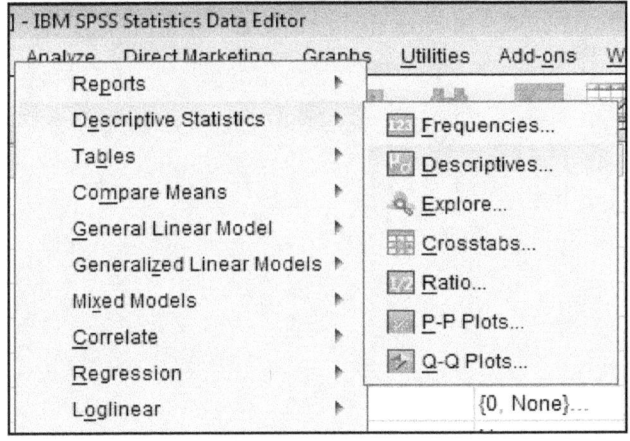

The following window appears:

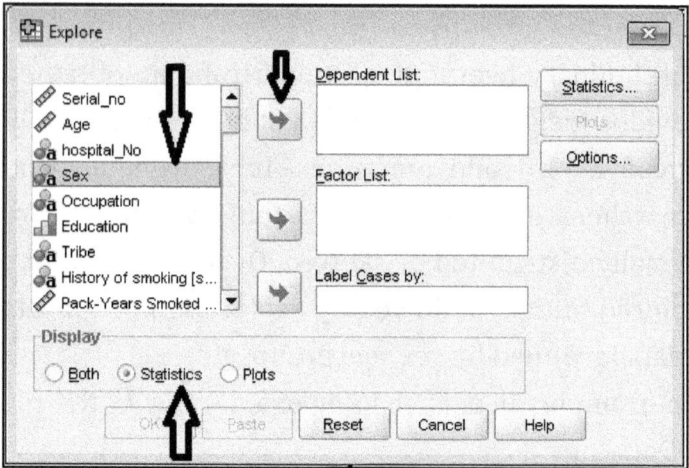

The "**Dependent List**" consists of the variables you wish to explore. Multiple variables can be added to or removed from the list using the arrow icons. The **Dependent List** accepts only numerical data.

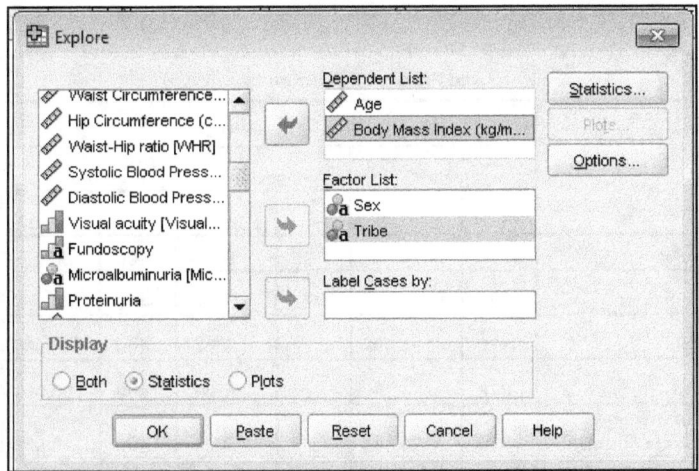

If you wish to have a bird's-eye view of a variable, leave the other fields empty. However, to explore sub-samples of the data (for example, age classified by gender), use the **Factor List** as well. Both lists can accept multiple variables to produce multiple tabular results.

By default, SPSS includes only cases that have complete data for all the variables in the lists.

Place all the scale variables in the file into the **Dependent Scale** and explore.

The results of **Explore** function displayed in the Viewer Window will show statistics only for the 378 cases (94.5%) with complete data.

Case Processing Summary						
	Cases					
	Valid		Missing		Total	
	N	Percent	N	Percent	N	Percent
Age	378	94.5%	22	5.5%	400	100.0%
Weight	378	94.5%	22	5.5%	400	100.0%
Height	378	94.5%	22	5.5%	400	100.0%
Waist-Hip ratio	378	94.5%	22	5.5%	400	100.0%
Systolic Blood Pressure (mmHg)	378	94.5%	22	5.5%	400	100.0%
Diastolic Blood Pressure (mmHg)	378	94.5%	22	5.5%	400	100.0%
PCV	378	94.5%	22	5.5%	400	100.0%
FBS	378	94.5%	22	5.5%	400	100.0%
HBA1c	378	94.5%	22	5.5%	400	100.0%
TC	378	94.5%	22	5.5%	400	100.0%
HDL	378	94.5%	22	5.5%	400	100.0%
LDL	378	94.5%	22	5.5%	400	100.0%
TGA	378	94.5%	22	5.5%	400	100.0%
Duration of Diabetes Mellitus	378	94.5%	22	5.5%	400	100.0%
Body Mass Index	378	94.5%	22	5.5%	400	100.0%
Diabetic Nephropathy	378	94.5%	22	5.5%	400	100.0%
TestCont	378	94.5%	22	5.5%	400	100.0%
Time till Death (days)	378	94.5%	22	5.5%	400	100.0%
Use of Cancidin	378	94.5%	22	5.5%	400	100.0%

Descriptives

			Statistic	Std. Error
Age	Mean		59.43	1.056
	95% Confidence Interval for Mean	Lower Bound	57.35	
		Upper Bound	61.50	
	5% Trimmed Mean		58.60	
	Median		58.00	
	Variance		421.529	
	Std. Deviation		20.531	
	Minimum		24	
	Maximum		400	
	Range		376	
	Interquartile Range		15	
	Skewness		12.135	.125
	Kurtosis		201.234	.250
Use of Cancidin	Mean		.6111	.02511
	95% Confidence Interval for Mean	Lower Bound	.5617	
		Upper Bound	.6605	
	5% Trimmed Mean		.6235	
	Median		1.0000	
	Variance		.238	
	Std. Deviation		.48814	
	Minimum		0.00	
	Maximum		1.00	
	Range		1.00	
	Interquartile Range		1.00	
	Skewness		-.458	.125
	Kurtosis		-1.800	.250

To ensure results for each variable appear independently, repeat the procedure but in the main dialog window, click the "**Options...**" button. In the dialog box that appears, select the option "**Exclude cases pairwise**". Click "**Continue**" to return to the main dialog window and then click "**OK**".

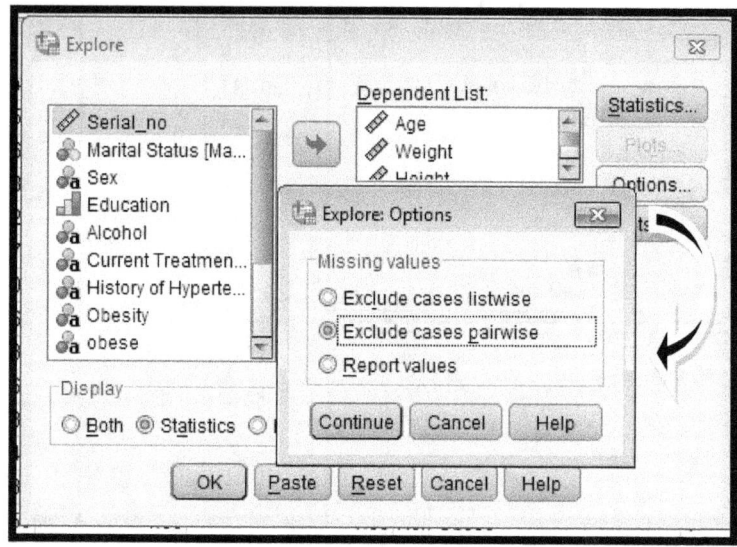

The case processing summary below shows that in this analysis, each variable is explored independently.

Case Processing Summary

	Cases					
	Valid		Missing		Total	
	N	Percent	N	Percent	N	Percent
Age	400	100.0%	0	0.0%	400	100.0%
Weight	400	100.0%	0	0.0%	400	100.0%
Height	394	98.5%	6	1.5%	400	100.0%
Waist-Hip ratio	400	100.0%	0	0.0%	400	100.0%
Systolic Blood Pressure (mmHg)	396	99.0%	4	1.0%	400	100.0%
Diastolic Blood Pressure (mmHg)	395	98.8%	5	1.3%	400	100.0%
PCV	393	98.3%	7	1.8%	400	100.0%
FBS	400	100.0%	0	0.0%	400	100.0%
HBA1c	400	100.0%	0	0.0%	400	100.0%
TC	400	100.0%	0	0.0%	400	100.0%
HDL	400	100.0%	0	0.0%	400	100.0%
LDL	400	100.0%	0	0.0%	400	100.0%
TGA	400	100.0%	0	0.0%	400	100.0%
Duration of Diabetes Mellitus	400	100.0%	0	0.0%	400	100.0%
Body Mass Index	400	100.0%	0	0.0%	400	100.0%
Diabetic Nephropathy	400	100.0%	0	0.0%	400	100.0%
TestCont	400	100.0%	0	0.0%	400	100.0%
Time till Death (days)	400	100.0%	0	0.0%	400	100.0%
Use of Cancidin	400	100.0%	0	0.0%	400	100.0%

Below are screenshots of **Age** explored in total and then **Age** explored using **Sex** as a factor.

Case Processing Summary

	Cases					
	Valid		Missing		Total	
	N	Percent	N	Percent	N	Percent
Age	400	100.0%	0	.0%	400	100.0%

Descriptives

			Statistic	Std. Error
Age	Mean		59.13	1.004
	95% Confidence Interval for Mean	Lower Bound	57.16	
		Upper Bound	61.11	
	5% Trimmed Mean		58.32	
	Median		58.00	
	Variance		403.148	
	Std. Deviation		20.079	
	Minimum		24	
	Maximum		400	
	Range		376	
	Interquartile Range		15	
	Skewness		12.293	.122
	Kurtosis		208.525	.243

In this exploration of **Age**, there are 400 cases analyzed with no missing cases. The maximum age is 400 years which is impossible, suggesting an error in data entry. Other statistics provided are invalid until the entry is corrected and data re-explored.

Skewness and Kurtosis are a measure of normality of the variable (discussed later).

After exploring **Age** between the Sexes, we find that there were 130 males and 270 females. The case with the wrongly-entered age is male (maximum age of 400 years, as opposed to maximum age of 82 years for the females). Therefore, the statistics for age of the females are still acceptable.

Search *myData2.sav* file for this male subject and correct his age to 40. Remember to save the file thereafter!

Case Processing Summary

		Cases					
		Valid		Missing		Total	
	Sex	N	Percent	N	Percent	N	Percent
Age	Male	130	100.0%	0	.0%	130	100.0%
	Female	270	100.0%	0	.0%	270	100.0%

				Statistic	Std. Error
	Female	Mean		57.96	.655
		95% Confidence Interval for Mean	Lower Bound	56.67	
			Upper Bound	59.25	
		5% Trimmed Mean		58.00	
		Median		58.00	
		Variance		115.891	
		Std. Deviation		10.765	
		Minimum		24	
		Maximum		82	
		Range		58	
		Interquartile Range		14	
		Skewness		-.026	.148
		Kurtosis		-.176	.295

Descriptives

	Sex			Statistic	Std. Error
Age	Male	Mean		61.57	2.769
		95% Confidence Interval for Mean	Lower Bound	56.09	
			Upper Bound	67.05	
		5% Trimmed Mean		58.97	
		Median		59.00	
		Variance		996.418	
		Std. Deviation		31.566	
		Minimum		39	
		Maximum		400	
		Range		361	
		Interquartile Range		15	
		Skewness		9.683	.212
		Kurtosis		104.259	.422

DESCRIBING DATA USING FREQUENCIES AND DESCRIPTIVES FUNCTION OF SPSS

Summary statistics for data include measures of dispersion and measures of central tendency.

Measures of central tendency include

- Mean: the average value obtained by dividing the sum of all the values by their number
- Median: the middle value when all the values are ranked from the minimum to the maximum. It corresponds to the 50th percentile.
- Mode: the most frequently occurring value

The measures of dispersion assess how widely the data is scattered around the mean. They include:

- Range: minimum value subtracted from maximum value
- Variance: obtained by (a) subtracting the mean from each data value (*deviation from the mean*); (b) squaring these differences; and then (c) getting the average of these squared deviations.
- Standard Deviation: square root of the variance

- Interquartile Range: difference between the 25th percentile and the 75th percentile.

  (When numerical data is ranked from minimum to maximum, the 25th percentile is the value that 25% of values of the variable fall below. Similarly, 75% of values are lower than the 75th percentile).

These measures are provided by default during the **Explore** procedure if the **Statistics** radio button is selected.

The **Descriptives** function allows you to select the statistics you desire. Apart from this feature and the presence of a "**Sum**" option, it is a watered-down version of the **Explore** function. To use it, Go to "**Analyze**" >> "**Descriptive Statistics**" >> "**Descriptives...**" Click the "**Statistics...**" button and tick the desired statistics.

The Frequencies function (**"Analyze"** >> **"Descriptive Statistics"** >> **"Frequencies..."**) is similar to Descriptives but includes Median, Mode, Percentiles, Quartiles and Cutoffs under its **"Statistics"**. It also provides the ability to produce tables and various charts.

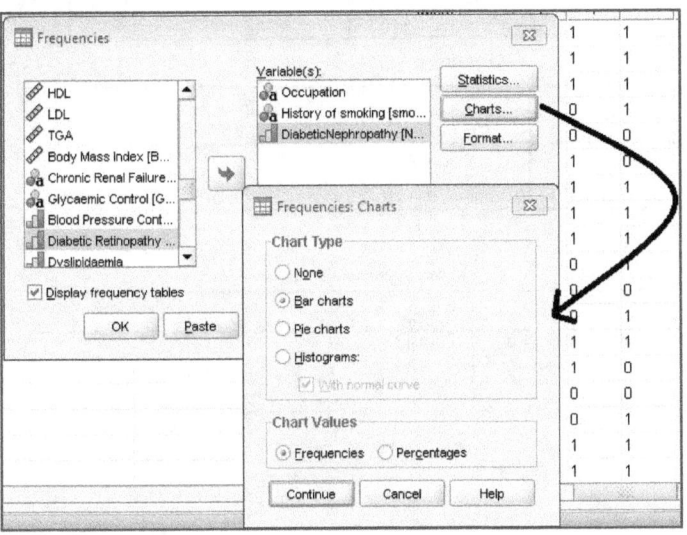

To avoid unnecessarily long tables, the **"Display frequency tables"** option is usually NOT selected when handling a numerical variable with many different values (for example, **Age**). The **"Display frequency tables"** option is particularly useful for categorical data. Results can be sorted in ascending or descending order of counts, percentages or values using the "**Format**" tab.

Below is the output of Frequencies as it appears in the Viewer window.

DiabeticNephropathy

		Frequency	Percent	Valid Percent	Cumulative Percent
Valid	Absent	191	47.8	47.8	47.8
	Present	209	52.2	52.2	100.0
	Total	400	100.0	100.0	

Statistics

		History of smoking	Diabetic Nephropathy
N	Valid	400	400
	Missing	0	0

History of smoking

		Frequency	Percent	Valid Percent	Cumulative Percent
Valid	No History Of Smoking	362	90.5	90.5	90.5
	Positive History of Smoking	38	9.5	9.5	100.0
	Total	400	100.0	100.0	

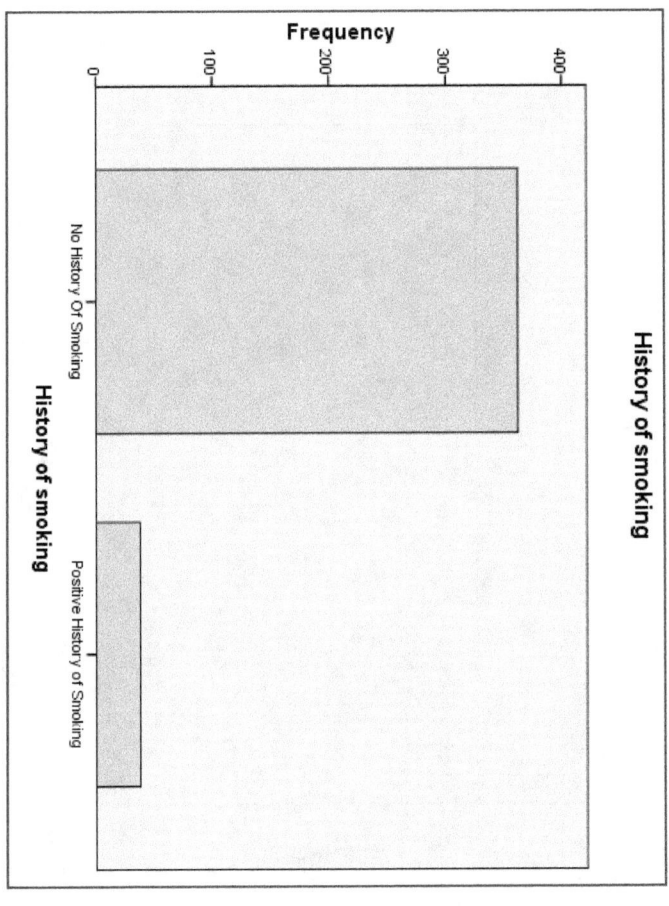

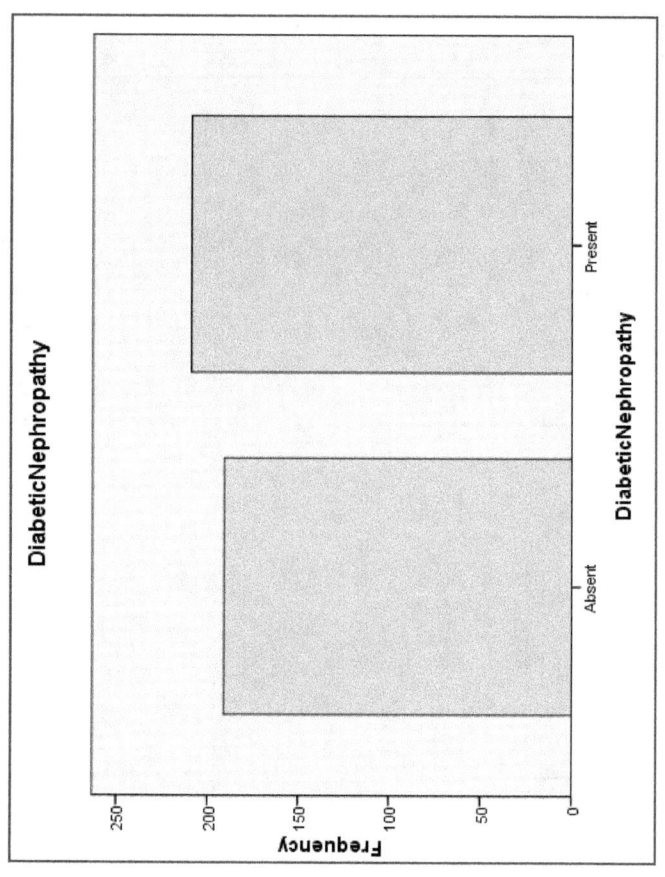

For variables defined in a multiple-response set, Frequencies is NOT used to produce tables. Instead, go to **"Analyze"** >> **"Multiple Response"** >> **"Frequencies"**. The typical output is shown below on the right:

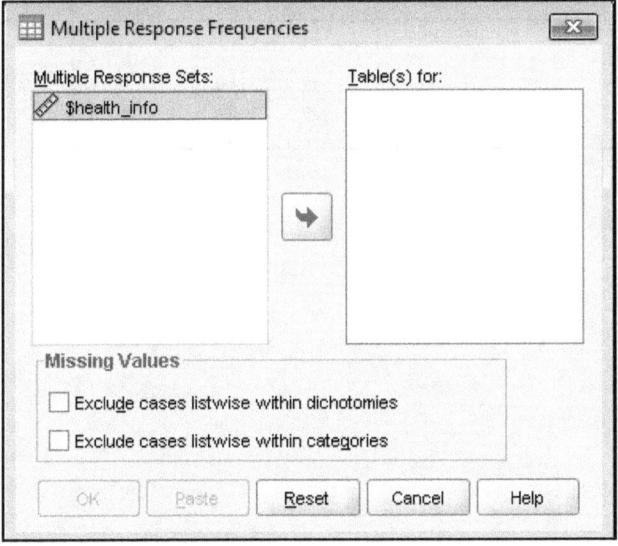

Case Summary

	Cases					
	Valid		Missing		Total	
	N	Percent	N	Percent	N	Percent
$health_info[a]	164	41.0%	236	59.0%	400	100.0%

a. Dichotomy group tabulated at value 0.

$health_info Frequencies

		Responses		Percent of Cases
		N	Percent	
Source of Health Information[a]	Newspapers	130	24.8%	79.3%
	TV	132	25.1%	80.5%
	Seminar	130	24.8%	79.3%
	Hospital	133	25.3%	81.1%
Total		525	100.0%	320.1%

a. Dichotomy group tabulated at value 0.

ASSESSING NORMALITY OF DISTRIBUTION OF CONTINUOUS DATA

The normal distribution or Gaussian distribution implies that as a continuous variable increases in value from its minimum to maximum values, the frequencies associated with these values increases to a peak corresponding to the mean (and mode as well as median) and declines thereafter. This produces a symmetrical bell-shaped histogram (the *normal* curve).

The mean, median and mode of a normally distributed variable are identical (or close in value). The standard deviation is usually less than half the value of the mean of a normal distributed variable.

To view the normal curve of the variable "Age" in our sample data, go to **"Analyze"** >> **"Descriptive Statistics"** >> "**Frequencies...**"

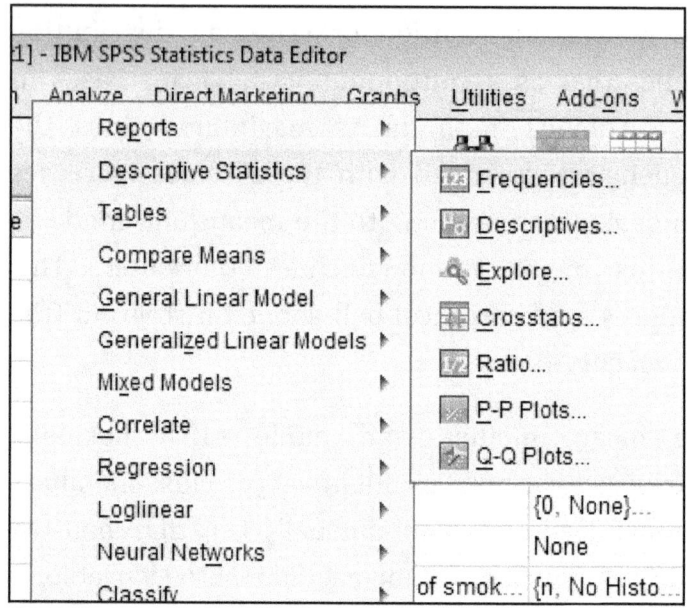

In the dialog window that appears,

1. Select "**Age**" into the field labelled "**Variable(s)**".
2. Click the button labelled "**Charts...**"
3. In the dialog box that appears, select the radio button labelled "**Histograms**" and tick the checkbox labelled "**Show normal curve on histogram**".
4. Click "**Continue**" to return to the main window. Ensure the checkbox labelled "**Display frequency tables**" is NOT checked.
5. Click "**OK**" to run the analysis.

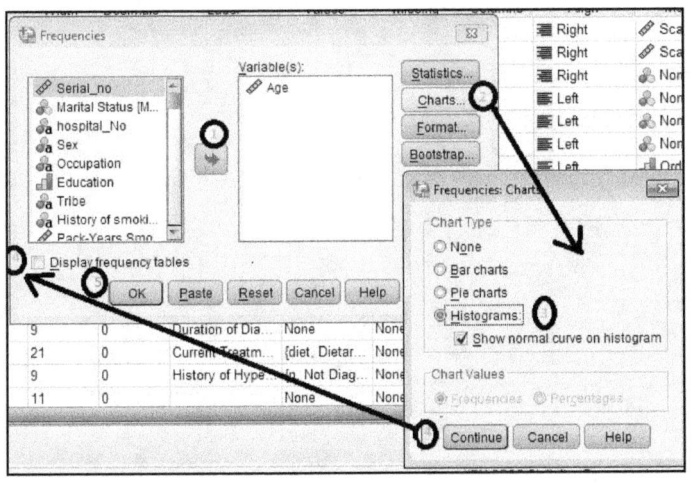

The resulting histogram and its normal curve is shown below.

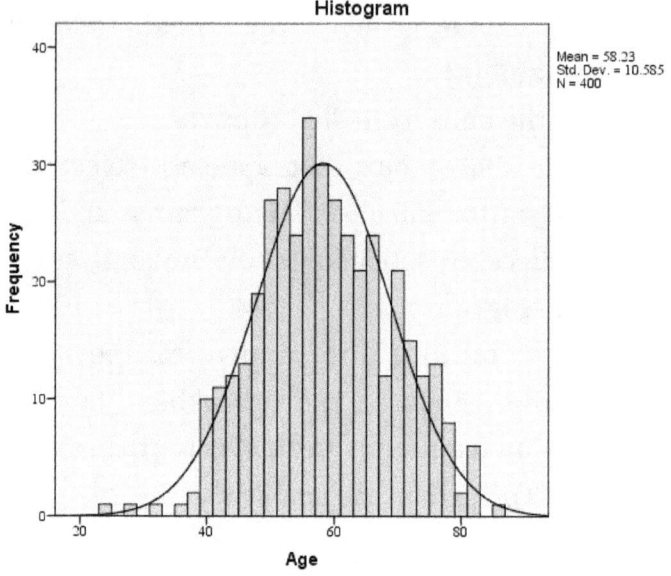

To objectively assess for normality of a variable, SPSS provides several options. The most comprehensive method is the "**Explore**" function.

Using our sample data, let us analyze the normality of distribution of "**Age**".

Go to "**Analyze**" >> "**Descriptive Statistics**" >> "**Explore**"

In the dialog window that appears, place the variable "**Age**" in the field marked "**Dependent List:**"

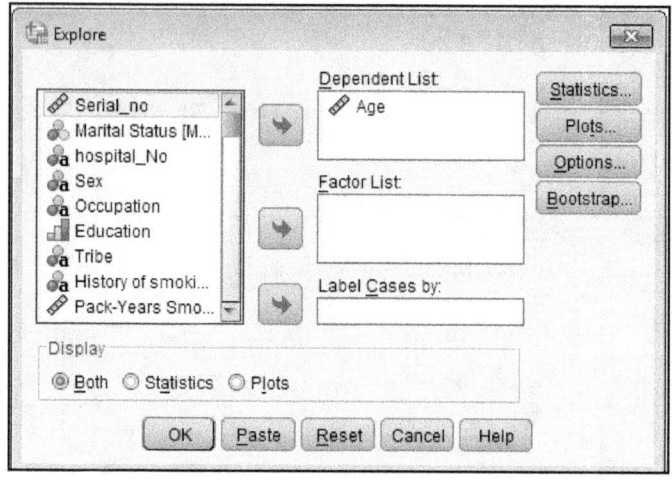

Click the button marked "**Plots...**" In the dialog box that appears, select "**Normality plots with tests**". Click "**Continue**" to return to the main window.

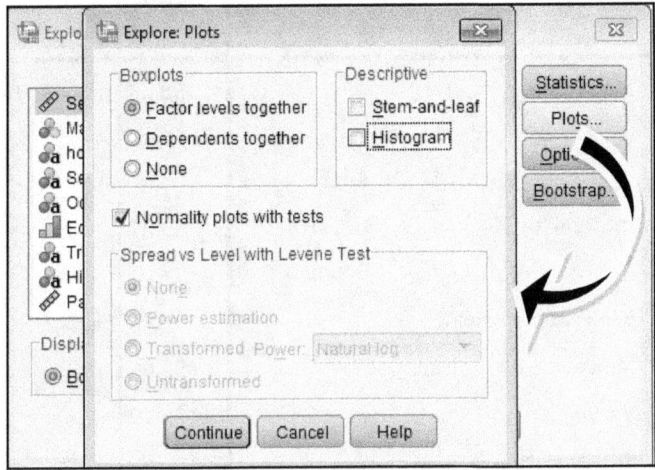

In the main window, ensure that the radio button labelled "**Both**" is selected. Click "**OK**" to run the analysis.

The results of the analysis are displayed below:

Descriptives

			Statistic	Std. Error
Age	Mean		58.23	.529
	95% Confidence Interval for Mean	Lower Bound	57.19	
		Upper Bound	59.27	
	5% Trimmed Mean		58.22	
	Median		58.00	
	Variance		112.049	
	Std. Deviation		10.585	
	Minimum		24	
	Maximum		85	
	Range		61	
	Interquartile Range		14	
	Skewness		.029	.122
	Kurtosis		-.283	.243

Note the closeness in value of the mean, the 95% confidence intervals for the mean, the 5% trimmed mean and the median. The standard deviation here is about one-sixth of the value of the mean.

Tests of Normality

	Kolmogorov-Smirnov[a]			Shapiro-Wilk		
	Statistic	df	Sig.	Statistic	df	Sig.
Age	.045	400	.051	.993	400	.065

a. Lilliefors Significance Correction

Kurtosis values and *skewness* values between -1 and +1 are expected for a normal distribution.

The *Kolmogorov-Smirnov (K-S) test* and *Shapiro-Wilk* test also tell us the data is normally distributed because their significance values are greater than or equal than 0.05. This may seem weird since most other tests give desirable results when their significance values are less than 0.05).

Normal Q-Q Plot of Age

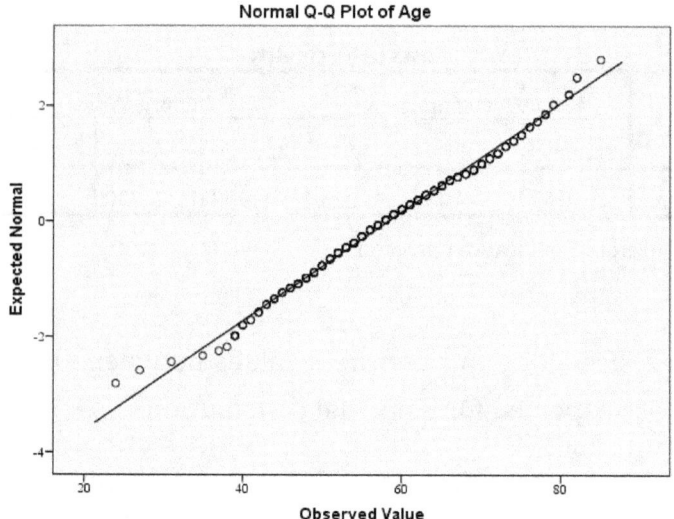

The closer the Q-Q plot is to the straight line, the greater the normality of the data. The plot above shows some deviation from normality at both extremes of data values.

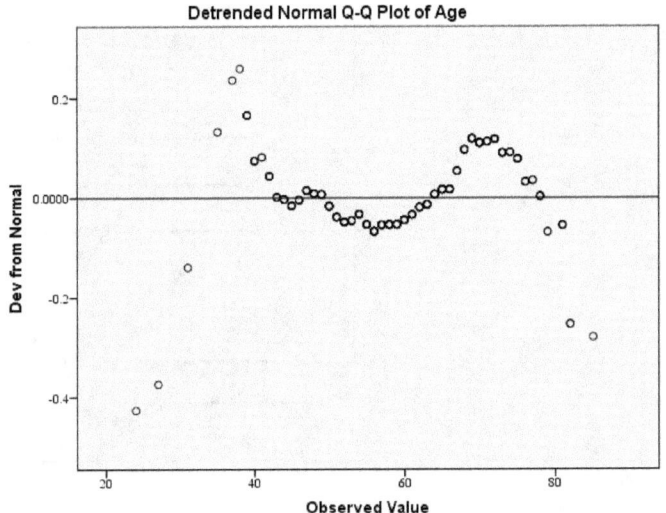

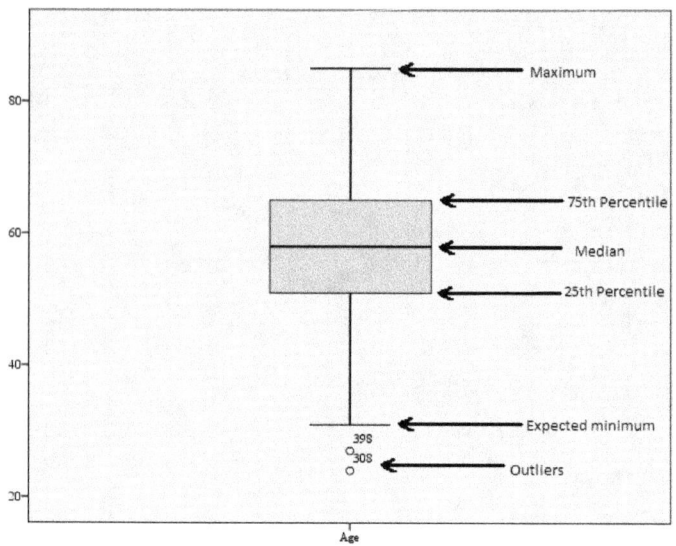

The boxplot shows that the 25th and 75th percentiles are about the same distance from the median. Although the ends of the whiskers are also symmetrical from the median, there are two outliers with case numbers 308 and 398 in the data.

CROSSTABULATIONS AND CORRELATIONS

Here we enter the grey areas between *describing* data and *analyzing* data.

Cross tables (or *contingency tables*) are used to show the relationship between a categorical variable and another variable which may be categorical. Several statistics can be calculated for this association but the most commonly used is the *Pearson chi-square*, often simply referred to as chi-square (χ^2).

Correlation is used to assess the relationship between a numerical variable and another numerical variable. Correlation will be discussed along with regression in a subsequent chapter.

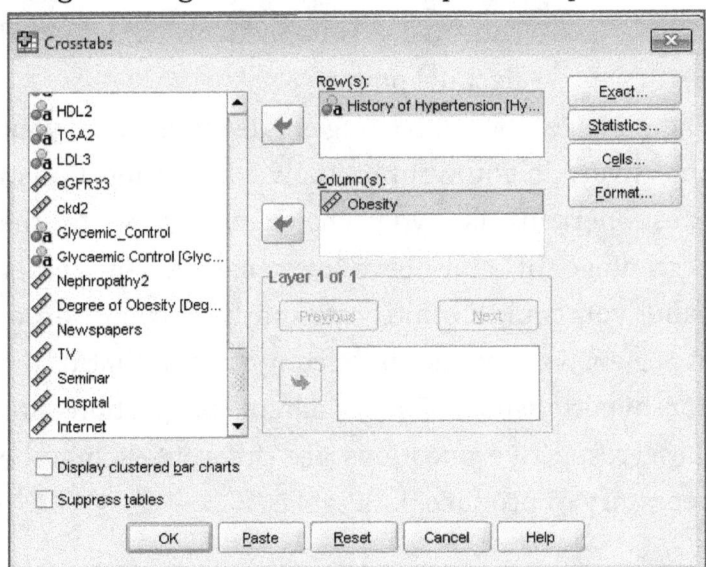

An example of the use of cross-tables for a purely descriptive purpose is when to compare proportions for example, we want to know what percentage of a population are obese AND hypertensive, obese but NOT hypertensive, hypertensive but NOT obese, and neither hypertensive nor obese.

Go to "**Analyze**" >> "**Descriptive Statistics**" >> "**Crosstabs…**" Select the variables to be cross tabulated in the rows and columns. The "Format…" tab determines if the rows of the table will be shown in ascending or descending order. The "**Cells…**" tabs determines what values and percentages are shown in the cells of the output table. This may be

- Observed and/or expected values
- Row, column and/or total percentages.
- Truncated, rounded or unadjusted decimal points.

The choice of percentages depends on your objectives: do you want to know what proportion of obese persons is hypertensive compared to the percentage of non-obese persons? On the other hand, you could be interested in the proportion of hypertensives that is obese compared to that of the non-hypertensives. The resultant proportions for these respective questions are different. It may be necessary to produce both tables.

The screen-captures below show typical selections and the resultant tables.

Case Processing Summary

	Cases					
	Valid		Missing		Total	
	N	Percent	N	Percent	N	Percent
History of Hypertension * Obesity	400	100.0%	0	.0%	400	100.0%

History of Hypertension * Obesity Crosstabulation

			Obesity		Total
			Obese	Not Obese	
History of Hypertension	Not Diagnosed Hypertensive	Count	83	136	219
		% within History of Hypertension	37.9%	62.1%	100.0%
	Diagnosed Hypertensive	Count	50	131	181
		% within History of Hypertension	27.6%	72.4%	100.0%
Total		Count	133	267	400
		% within History of Hypertension	33.2%	66.8%	100.0%

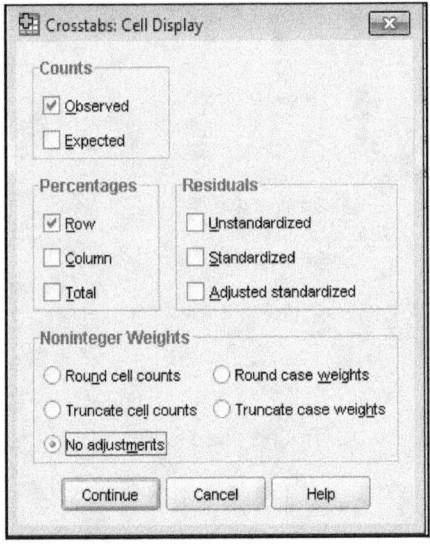

In this case, 27.6% of the hypertensive subjects were obese, compared to 37.9% of the non-hypertensive subjects. Overall, 33.2% of the subjects were obese.

If we create another table using the same variables but showing column percentages instead, we deduce that 37.6% of the obese subjects were hypertensive compared to 49.1% of the non-obese subjects. Overall, 45.2% of the subjects were hypertensive.

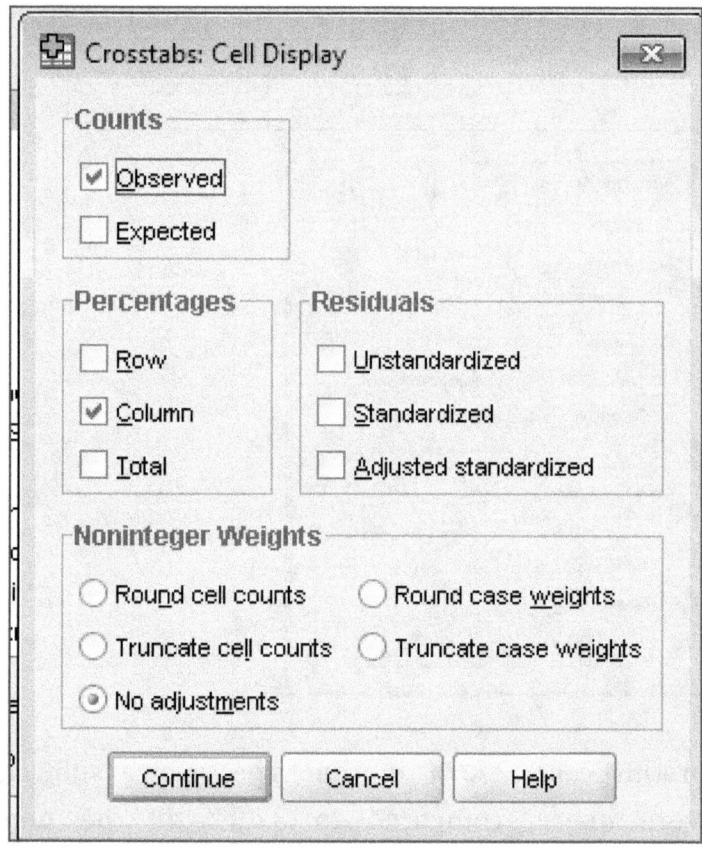

Case Processing Summary

	Cases					
	Valid		Missing		Total	
	N	Percent	N	Percent	N	Percent
History of Hypertension * Obesity	400	100.0%	0	.0%	400	100.0%

History of Hypertension * Obesity Crosstabulation

			Obesity		Total
			Obese	Not Obese	
History of Hypertension	Not Diagnosed Hypertensive	Count	83	136	219
		% within Obesity	62.4%	50.9%	54.8%
	Diagnosed Hypertensive	Count	50	131	181
		% within Obesity	37.6%	49.1%	45.2%
Total		Count	133	267	400
		% within Obesity	100.0%	100.0%	100.0%

The function of the "**Statistics...**" tab will be discussed during the section on hypothesis testing.

Output of cross-tabulation can also be displayed as clustered bar charts (select the check box "**Display clustered bar charts**". The column (independent) variable forms the category (horizontal) axis.

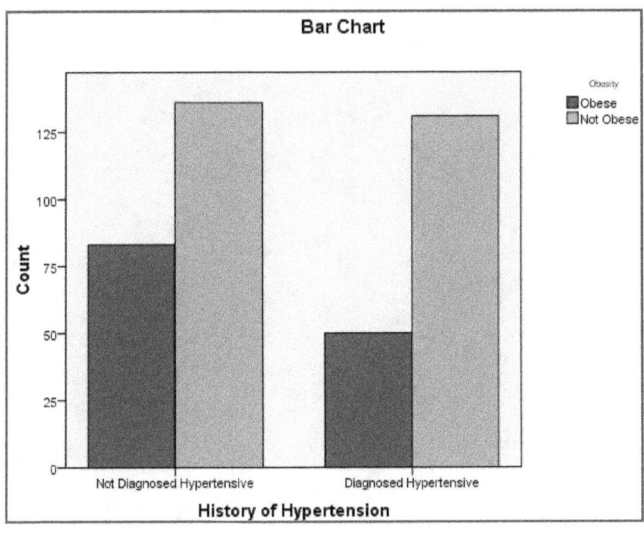

INFERENTIAL STATISTICS

Inferential statistics allow us to analyze a sample and extrapolate our conclusions to the population with a reasonable amount of certainty.

BASIC CONCEPTS

Significance: This is the most... significant term in inferential statistics.

Let us assume we have two groups of adults placed on drugs A and B respectively for blood glucose control. After a month of therapy, the mean reduction of fasting blood glucose levels for subjects on drug A is 30mg/dl while for those on drug B, it is 33mg/dl. Is this enough reason to assume that drug A is more effective than drug B? Is the difference significant?

The research hypothesis (also called the alternate hypothesis) is that there is a difference between the mean reductions comparing drug A against drug B, AND that this difference is not a chance occurrence.

The null hypothesis is the opposite of this: there is no difference OR any difference obtained can be explained by chance.

The aim of hypothesis testing is to prove or disprove this null hypothesis to a reasonable degree of certainty. For most studies, this degree of certainty is set at 95% (this is also the default setting in SPSS). Therefore, if the probability of the null hypothesis (p) is less than 5% (0.05), the research hypothesis has over 95% probability of being true and is accepted. *P* is also called the type 1 (or α) error.

Statistical significance implies that a difference or association observed is NOT likely to be from chance and hence it should be accepted. However, a statistically significant difference may not be *practically* significant in the real world. In this example, an extra 3mg/dl reduction in fasting blood glucose may not be so important in clinical practice.

In cases where a large difference is obtained but is found to be statistically insignificant, it may suggest an inappropriately small sample size. A larger study should be considered.

Association: implies a relationship between two or more variables. The strength and statistical significance of this relationship can be assessed using appropriate tests.

Parametric tests: tests designed to be used on normally distributed continuous variables. These are the most commonly used tests for hypothesis testing. *Non-parametric tests* are available for ordinal data or data that is not normally distributed but they are not as reliable. However, if the sample size is greater than 30, normality can be assumed and parametric tests used. We shall emphasize the parametric tests.

CHOOSING AN APPROPRIATE TEST

In choosing an appropriate test, the questions to ask are:

1. Am I comparing groups for *differences* (chi-square, t-test, ANOVA) or am I checking for *associations* (Chi-square, correlation, odds ratio, relative risk, regression)?

2. Are the predictor and outcome variables nominal, binary, ordinal or scale?

3. How many groups are going to be compared?

4. How many possible outcomes will be compared?

Objective: COMPARISON

No of groups	Data Type	Test
1	Scale (compared against a known population parameter)	One-Sample t-test
1	Ordinal/Non-Parametric (compared against a known population parameter)	Sign test
2	Scale (independent groups)	Independent samples t-test
2	Ordinal/Non-Parametric (independent groups)	Mann-Whitney U test
2	Scale (paired, matched or repeated data)	Paired-samples t-test
2	Ordinal/Non-Parametric (paired, matched or repeated data)	Wilcoxon Signed-Ranks test
2	Nominal (paired, matched or repeated data)	McNemar test
3+	Scale (Independent groups)	ANOVA
3+	Ordinal/Non-Parametric (independent groups)	Kruskal-Wallis test
3+	Scale (paired, matched or repeated data)	Repeated-measures ANOVA
3+	Ordinal/Non-Parametric (paired, matched or repeated data)	Friedman's test
Any no.	Nominal	Chi-Square

Objective: ASSOCIATION

Predictor Variable	No.	Outcome Variable	Test
Scale	1	Scale	Pearson Correlation /Linear Regression
Ordinal (or Scale)	*1*	*Ordinal (or Scale)*	*Spearman Rank Correlation*
Binary	1	Binary	Odds Ratio/ Relative Risk/ Chi-Square
Scale, binary	1	Binary	Simple Logistic regression
Binary, ordinal, scale, nominal	2+	Scale	Multiple linear regression
Binary, ordinal, scale, nominal	2+	Binary	Multiple logistic regression
Nominal	2+	Nominal	Chi-Square

THE ONE-SAMPLE T-TEST

This is used to compare the mean of a variable in a sample to that of the population of interest. An example of a pertinent research question is *"Does the mean age of our sample differ significantly from the national average?"* Assuming the national mean is 25 years, go to **"Analyze"** >> **"Compare Means"** >> **"One-Sample T test..."**

In the dialog window that appears, select the variable and then type in **25** in the number field labelled **"Test Value"**. The output is shown below:

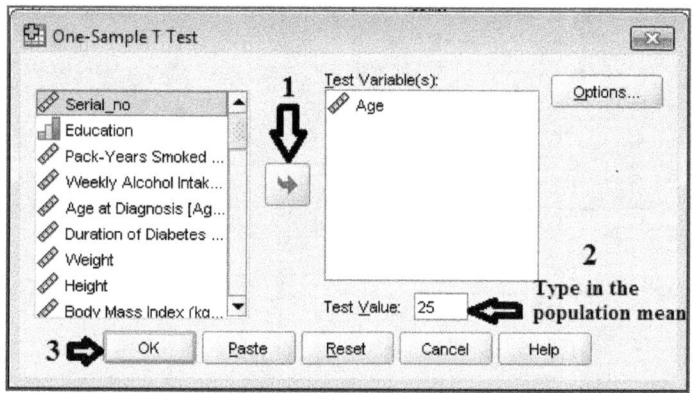

One-Sample Statistics

	N	Mean	Std. Deviation	Std. Error Mean
Age	400	59.13	20.079	1.004

One-Sample Test

	Test Value = 25				
				95% Confidence Interval of the Difference	
	t	df	Sig. (2-tailed)	Mean Difference	Lower	Upper
Age	33.999	399	.000	34.132	32.16	36.11

t value ⇐

p value ⇐

⇐ =Sample mean - population mean

In this case, the sample mean (59.13 years) is greater than the population mean (25 years) and the difference is statistically significant ($p<0.01$).

The mean difference is 34.13 years with a 95% chance that its true value lies between 32.16 years and 36.11 years.

If the lower confidence interval has a different sign from the upper confidence interval, it also implies that the difference is not significant.

THE INDEPENDENT-SAMPLES T TEST

The independent-samples t test is used to compare the means of two independent groups.

This t-test is either *two-tailed* (non-directional, A not equal to B) or *one-tailed* (directional, A>B or B>A). However, two-tailed tests are generally preferred.

An example of such a research question for a one-tailed test would be *"Do the male subjects have higher HDL cholesterol levels than the females?"*

A two-tailed test would answer the question *"Is there any difference in HDL cholesterol levels between males and females?"* The direction of the difference can be ascertained from the sign of the mean difference (positive or negative).

To carry out a two-tailed independent samples T test, go to **"Analyze"** >> **"Compare Means"** >> **"Independent-Samples T test..."**

In the dialog window, select the outcome variable(s) and choose the grouping variable.

Select **"Define Groups..."** and type in the value codes of the grouping variable.

Click **"Continue"** to close this dialog window and then click **"OK"**.

The screenshots and output tables are shown below:

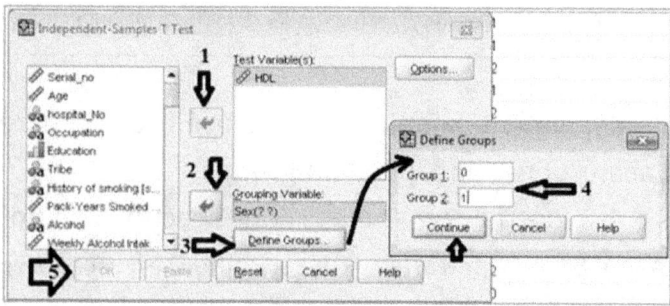

Group Statistics					
	Sex	N	Mean	Std. Deviation	Std. Error Mean
HDL	Male	130	51.35	17.671	1.550
	Female	270	49.32	18.026	1.097

Independent Samples Test

		Levene's Test for Equality of Variances		t-test for Equality of Means					95% Confidence Interval of the Difference	
		F	Sig.	t	df	Sig. (2-tailed)	Mean Difference	Std Error Difference	Lower	Upper
HDL	Equal variances assumed	.024	.876	1.066	398	.287	2.038	1.912	-1.721	5.798
	Equal variances not assumed			1.074	259.413	.284	2.038	1.899	-1.701	5.778

The output shows the summary statistics for each group, showing a higher mean for the males than for the females.

Note also the presence of two rows with two different t-test results in the second output table. The upper row gives a t-value and a p-value with the assumption that the standard deviations (and variances) of the two groups are equal. A preliminary test (Levene's test) is used to test if the variances are different. If the significance value of this Levene's test is less than 0.05, then the assumption is that the variances are not equal and the associated t-test is used.

In this example, the p value for the Levene's test is >0.05 (p=0.876), so equal variances are assumed.

The associated t-value is 1.066, with p=0.287. Therefore, there is no significant difference between the mean HDL levels between males and females.

The rule on inequality of signs of the confidence intervals also applies here. Note that the lower confidence interval is negative while the upper confidence interval is positive, also implying a non-significant difference.

THE PAIRED-SAMPLES T- TEST

The paired-samples t-test is used to compare the means of a pair of related samples. Such data may be from matched subjects or from repeated observations of the same set of subjects.

An example of such data is systolic blood pressure measured in cases compared to matched controls, or systolic blood pressure measured in subjects before and after use of a drug.

Go to **"Analyze"** >> **"Compare Means"** >> **"Paired-Samples T Test..."** Select the pairs.

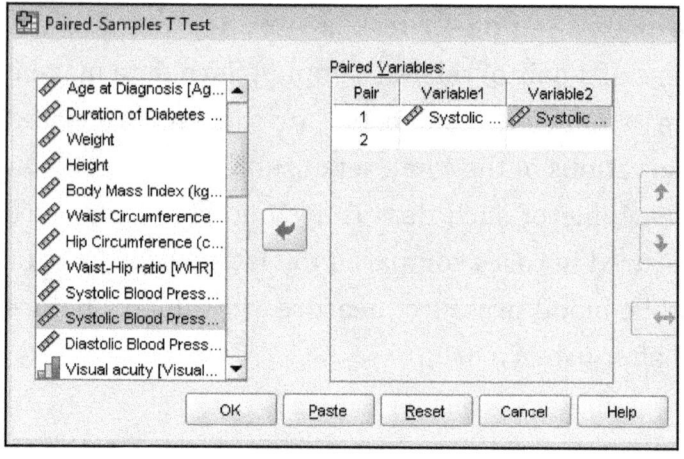

Paired Samples Statistics

	Mean	N	Std. Deviation	Std. Error Mean
Pair 1	Systolic Blood Pressure of Cases (mmHg) 141.04	400	20.861	1.043
	Systolic Blood Pressure of Controls (mmHg) 147.6100	400	17.77816	.88891

Paired Samples Correlations

	N	Correlation	Sig.
Pair 1	Systolic Blood Pressure of Cases (mmHg) & Systolic Blood Pressure of Controls (mmHg) 400	.115	.022

Paired Samples Test

	Paired Differences			95% Confidence Interval of the Difference				
	Mean	Std. Deviation	Std. Error Mean	Lower	Upper	t	df	Sig. (2-tailed)
Pair 1	Systolic Blood Pressure of Cases (mmHg) - Systolic Blood Pressure of Controls (mmHg) -6.57125	25.80890	1.29045	-9.10817	-4.03433	-5.092	399	.000

The results show the T test result as well as a correlation (which we shall ignore).

The T test shows a t-value of -5.092 and p-value<0.001, which is significant. The values are negative because the controls had higher mean systolic blood pressure readings than the cases.

Note that both limits of the confidence interval are also negative. This is another method of checking statistical significance with t-test. If the signs of the limits of the confidence interval are in the same direction (both positive/plus or both negative/minus), the t-test is significant.

ONE-WAY ANALYSIS OF VARIANCE

Assuming we intend to compare the means of three groups, what do we do? A possibility would be to compare each group against each of the other two, resulting in three independent-samples t-test results (A vs. B, A vs. C, and B vs. C). The problem with this is that for every extra comparison performed on the same predictor variable in the same sample, the Type 1 error increases proportionately. In this example, the α-error increases to 0.15 (0.05 x 3).

Analysis of variance allows you to compare 3 or more groups for significant differences without loss of analytical power. If a significant difference exists, a *post-hoc test* may be performed to detect the particular groups that differ significantly. There are several of these post-hoc tests; in clinical medicine, Bonferroni test is often used while Duncan test is usually selected in public health studies.

To do a one-way ANOVA, go to **"Analyze"** >> **"Compare Means"** >> **"One-Way ANOVA..."**

Step 1: Choose the dependent continuous variable(s) to be analyzed.

Step 2: Select the **Factor** variable (similar to grouping variable used for independent samples t-tests).

Step 3: Choose the desired post-hoc test(s).

Example of research question: Do packed cell volume levels differ significantly among diabetics on different treatment regimens (diet alone, oral drugs alone, insulin alone, oral drugs and insulin)?

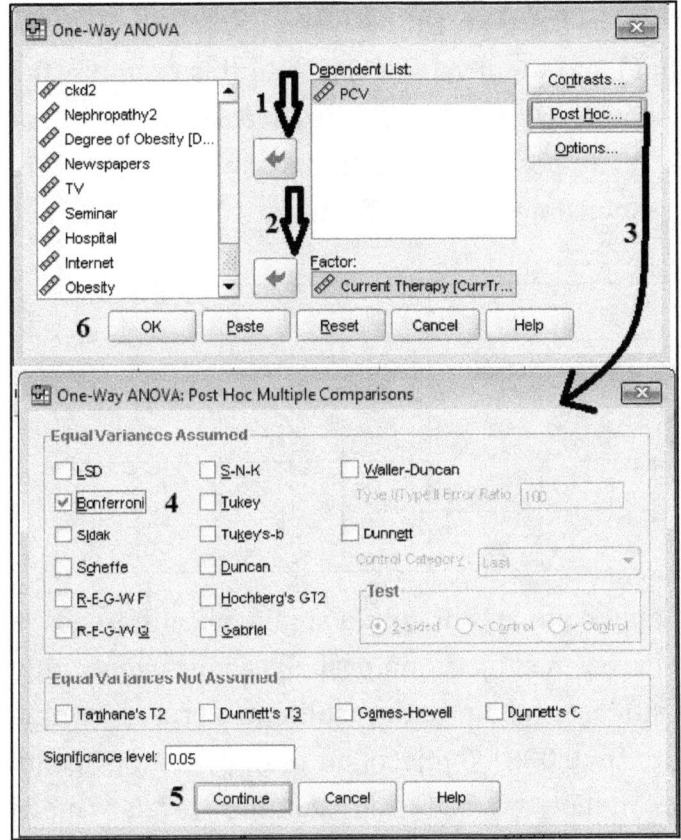

The following is the result of the ANOVA:

ANOVA

PCV

	Sum of Squares	df	Mean Square	F	Sig.
Between Groups	281.970	3	93.990	4.766	.003
Within Groups	7809.937	396	19.722		
Total	8091.906	399			

The relevant statistics here are the F value, the two

degrees of freedom (df between groups, and within groups) and the *p* value ("Sig."). In this example, the one-way ANOVA is significant *(p=0.634)* so we shall look at the post-hoc tests to determine which pairs are responsible for the difference.

Multiple Comparisons

PCV
Bonferroni

(I) Current Therapy	(J) Current Therapy	Mean Difference (I-J)	Std. Error	Sig.	95% Confidence Interval	
					Lower Bound	Upper Bound
Diet only	Insulin	3.887	1.650	.114	-.49	8.26
	Insulin and oral glycaemic agents	2.707	1.349	.273	-.87	6.28
	Oral glycaemic agents	.821	1.212	1.000	-2.39	4.04
Insulin	Diet only	-3.887	1.650	.114	-8.26	.49
	Insulin and oral glycaemic agents	-1.180	1.314	1.000	-4.66	2.30
	Oral glycaemic agents	-3.065	1.173	.056	-6.18	.04
Insulin and oral glycaemic agents	Diet only	-2.707	1.349	.273	-6.28	.87
	Insulin	1.180	1.314	1.000	-2.30	4.66
	Oral glycaemic agents	-1.885*	.687	.038	-3.71	-.06
Oral glycaemic agents	Diet only	-.621	1.212	1.000	-4.04	2.39
	Insulin	3.065	1.173	.056	-.04	6.18
	Insulin and oral glycaemic agents	1.885*	.687	.038	.06	3.71

*. The mean difference is significant at the 0.05 level.

The only comparison with a significant difference is that between subjects on oral glycaemic agents only and subjects taking both insulin and oral glycaemic agents (*p*=0.038). Subjects on oral glycaemic agents have higher packed cell volume levels (mean difference =1.885, CI= 0.06, 3.71). Do you think this statistically significant difference is *clinically* significant?

CHI-SQUARE AND RISK

The chi-square is a versatile non-parametric test that can be applied to virtually every data analysis (except multiple-response data). It can be applied to numerical data after it has been categorized for example, **BMI** can be recoded into "**underweight**", "**normal**", "**underweight**" and "**obese**" before being compared between males and females. This will produce a cross-table with 2 columns and 4 rows.

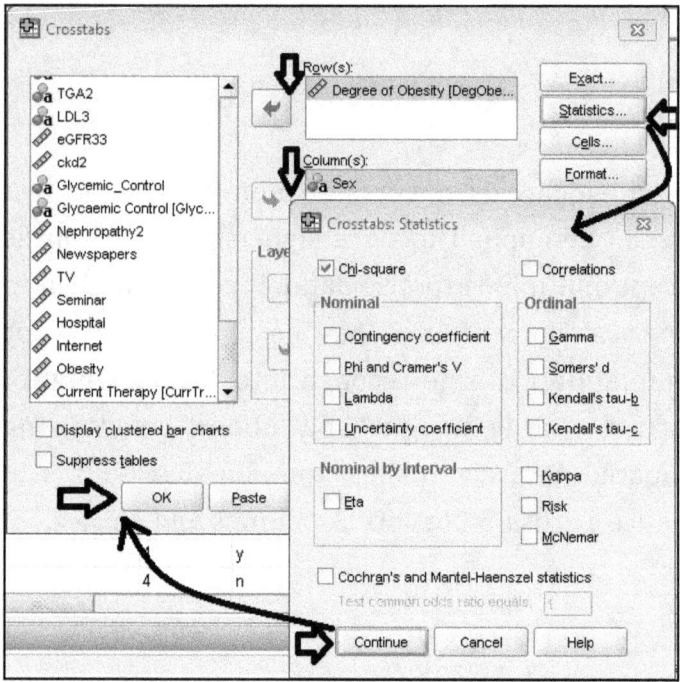

To do this, go to **"Analyze"** >> **"Descriptive Statistics"** >> **"Crosstabs..."**. Move the dependent variable **"Degree of Obesity"** into the "**Column(s)**" field and the independent variable "**Sex**" into the "**Row(s)**" field. Click the "**Statistics...**" tab and tick the "**Chi-square**" checkbox. Click **"Continue"** and then "**OK**".

The output is shown below.

Degree of Obesity * Sex Crosstabulation					
			Sex		
			Male	Female	Total
Degree of Obesity	Underweight	Count	1	2	3
		% within Sex	.8%	.7%	.8%
	Normal	Count	41	79	120
		% within Sex	31.5%	29.3%	30.0%
	Overweight	Count	47	97	144
		% within Sex	36.2%	35.9%	36.0%
	Obese	Count	41	92	133
		% within Sex	31.5%	34.1%	33.2%
Total		Count	130	270	400
		% within Sex	100.0%	100.0%	100.0%

Chi-Square Tests

	Value	df	Asymp. Sig. (2-sided)
Pearson Chi-Square	.324ᵃ	3	.955
Likelihood Ratio	.324	3	.955
N of Valid Cases	400		

a. 2 cells (25.0%) have expected count less than 5. The minimum expected count is .98.

Note that ideally each cell's *expected value* should be more than 5. (This is calculated for each cell as row total x column total divided by the grand total). The options are to use the chi-square as it is (provided less than 20% of cells are affected) or to recode the data so that those rows with small values merge with other rows. In this case, there is no significant association between sex and BMI as Pearson $\chi 2$ is 0.324 with 3 degrees of freedom and a p of 0.955.

Fisher's exact chi-square and chi-square with Yates' (continuity) correction are applicable to only tables with two rows and two columns (2 x 2 tables) only. They are used when cells have expected values less than 5.

Risk is also calculated only for 2 x 2 tables. By convention, when calculating risk, the independent variables (*exposure, risk factor*) occupy the rows while the dependent variables (*reaction, outcome*) are placed in the columns. To obtain risk statistics, go through the steps for carrying out a Chi-Square but also tick "**Risk**" under the "**Statistics…**" dialog.

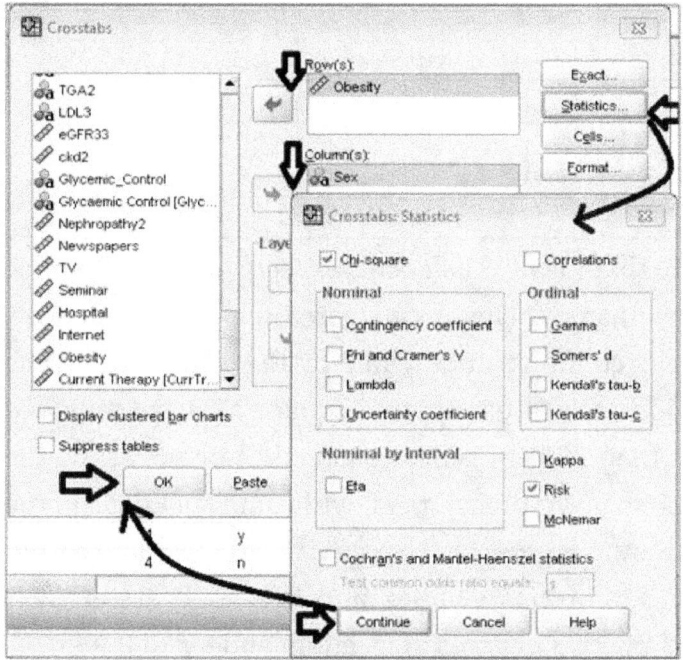

In the data used to produce the result below, male sex was coded as 1 and female sex as 0. In this example, male sex is the risk factor we are looking at.

Risk Estimate

	Value	95% Confidence Interval	
		Lower	Upper
Odds Ratio for Sex (Female / Male)	.759	.416	1.387
For cohort Obesity = Absent	.907	.723	1.138
For cohort Obesity = Present	1.195	.820	1.741
N of Valid Cases	400		

For prospective (cohort) studies, use the *Relative Risk* (RR). It assesses the *probability* of the outcome in the exposed group against the probability in the unexposed group.

The *Odds Ratio* (OR) can be used for many types of studies. However it is difficult to interprete directly as it is a ratio of ratios. The *odds* of an event is the ratio of the probability of the event happening to the probability of the event *not* happening. The probability of getting "heads" after a single toss of a coin is 50% or 0.5. The odds of getting a "heads" is

50/50 which is 1.

The *odds ratio* is the ratio of the *odds* (**not** *probability*) of getting the outcome in the exposed group to the odds of getting the outcome in the unexposed group.

Dicey, ain't it?

For very *unlikely* outcomes, RR and OR become nearly equal.

Assuming the result above is from a retrospective study, the odds ratio (male sex as risk factor for obesity) is 0.759 with confidence intervals of 0.416 and 1.387. The odds ratio is less than 1 implying a *protective* association. Since the confidence intervals straddle 1, it is NOT significant as the risk could be either higher, lower or equal in males compared to females.

If the result is from a prospective study, the relative risk is obtained by getting the *reciprocal* of the value from the *third row* which shows as "**For cohort...=Present**". The value here is 1.195. The relative risk therefore is 1/1.195 which gives 0.837. Similarly, the confidence intervals shown (0.820, 1.741) need to be inverted, giving us 1.220 (upper limit) and 0.574 (lower limit).

CORRELATION AND REGRESSION

How does a change in one quantitative variable associate with changes in others? Can the value of a variable be predicted from other variables?

Correlation and regression attempt to answer these questions.

CORRELATION

This assesses if there is a linear association between two numerical variables (*bivariate correlation*). It also provides a numerical coefficient (**Pearson Product-Moment Coefficient, r**) indicating the direction and strength of this association. Values range from -1 (*perfect inverse correlation*) to +1 (*perfect direct correlation*). Values close to +1 or -1 imply a strong linear relationship while values near zero suggest very weak linear association. The statistical significance ("**Sig. (2-tailed)**") of the association is also provided.

To assess for correlation between systolic blood pressure, HDL cholesterol and body mass index, go to "**Analyze**" >> "**Correlate**" >> "**Bivariate...**"

The results below show a statistically significant correlation between systolic blood pressure and HDL ($p=0.027$). However the r value is 0.111 suggesting that this is a weak linear association,

Please not that the absence of a *linear* correlation does not mean the variables do not have other forms of association (e.g. quadratic, u-shaped, or exponential relationships).

Correlation does NOT also imply causation!

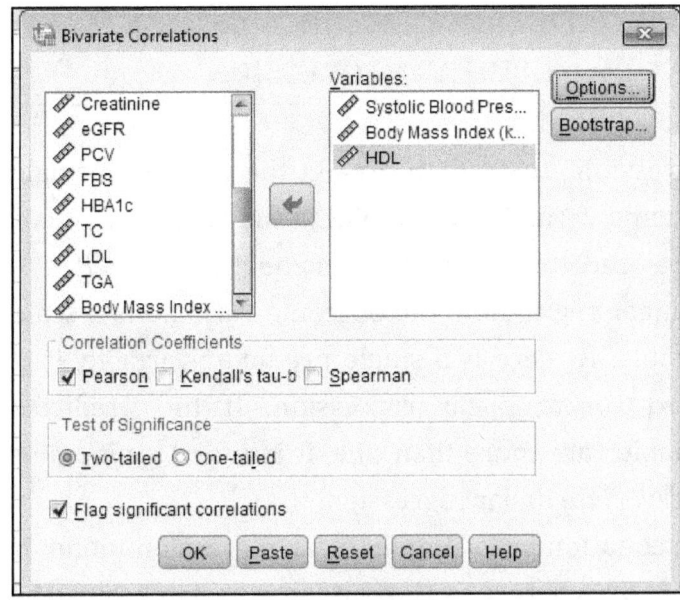

Correlations

		Systolic Blood Pressure of Cases (mmHg)	Body Mass Index (kg/m2)	HDL
Systolic Blood Pressure of Cases (mmHg)	Pearson Correlation	1	.059	.111*
	Sig. (2-tailed)		.235	.027
	N	400	400	400
Body Mass Index (kg/m2)	Pearson Correlation	.059	1	-.013
	Sig. (2-tailed)	.235		.802
	N	400	400	400
HDL	Pearson Correlation	.111*	-.013	1
	Sig. (2-tailed)	.027	.802	
	N	400	400	400

*. Correlation is significant at the 0.05 level (2-tailed).

LINEAR REGRESSION

Regression attempts to predict the value of a single outcome variable by combining the values of one or more predictor variables in an equation *(model)*.

In linear regression, the outcome variable is a scale variable. If there is a single predictor variable, it is called *simple* linear regression. If the predictor variables are more than one, it is called *multiple* or *multivariate* linear regression.

The equation for a simple linear regression model is $y=c + Bx$ where y is the outcome variable, x is the predictor variable, c is a constant and B is the **unstandardized** *regression coefficient* of x. The term *unstandardized* implies that B is related to the units of measurement of the variable. Hence if x is length measured in metres, B will be 100 times greater than B obtained when length is measured in centimeters.

β is the *beta weight* or *standardized regression coefficient* of x. it is independent of the units of measurement used. It is more useful when there are multiple predictor variables.

To carry out a simple linear regression between body mass index and waist-hip ratio, go to **"Analyze"** >> **"Regression"** >> **"Linear..."**

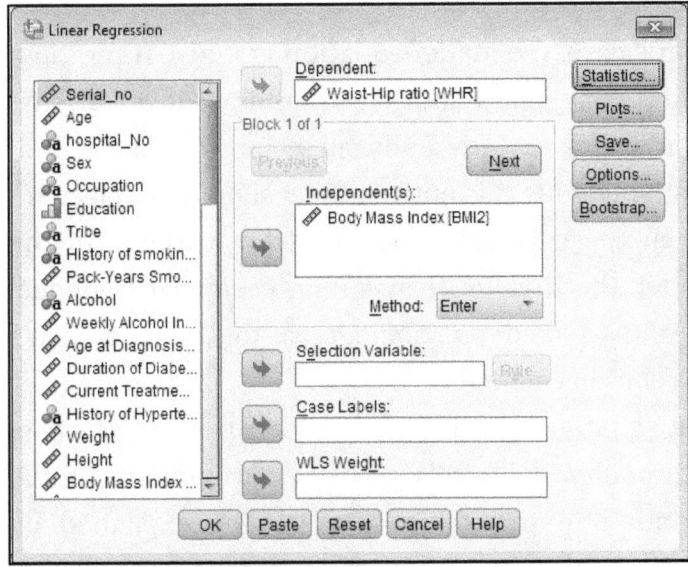

Model Summary[b]

Model	R	R Square	Adjusted R Square	Std. Error of the Estimate
1	.073[a]	.005	.003	.06307

a. Predictors: (Constant), Body Mass Index

b. Dependent Variable: Waist-Hip ratio

Coefficients[a]

Model		Unstandardized Coefficients		Standardized Coefficients	t	Sig.
		B	Std. Error	Beta		
1	(Constant)	.886	.012		73.134	.000
	Body Mass Index	.006	.004	.073	1.450	.148

a. Dependent Variable: Waist-Hip ratio

Residuals Statistics[a]

	Minimum	Maximum	Mean	Std. Deviation	N
Predicted Value	.8914	.9083	.9027	.00458	400
Residual	-.12472	.32521	.00000	.06299	400
Std. Predicted Value	-2.476	1.206	.000	1.000	400
Std. Residual	-1.978	5.156	.000	.999	400

a. Dependent Variable: Waist-Hip ratio

In this case, the model equation is Waist-Hip ratio = 0.006(Body mass index) + 0.886. However, the significance test for Body mass index shows a p of 0.148 meaning this entire model should be rejected.

When there is only one predictor variable in the model, the R value (*correlation coefficient*) is equal to the β and is analogous to the Pearson's correlation coefficient.

The square of the R (R^2) is called the *coefficient of determination*. It tells us what proportion of the variation in the value of y can be explained by changes in x. In the example shown above, R^2 is 0.005, implying that x accounts for only 0.5% of the variation in y.

The equation for multiple linear regression for n independent variables is similar: $y = B_1x_1 + B_2x_2 + B_3x_3 + ...B_nx_n + c$. However other considerations here include:

- F test (ANOVA test): to determine how well the overall equation fits. A higher F value suggests better fit

- Method of variable selection and entry into the model. There are several methods but for simplicity, use the default method "Entry".

To perform multiple linear regression, follow the same steps listed for simple linear regression but include all other predictor variables in the field **"Independent(s)"**

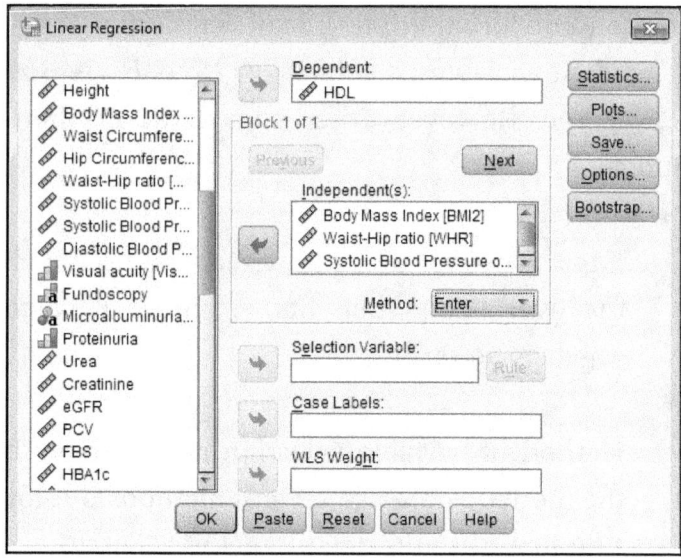

Model Summary[b]

Model	R	R Square	Adjusted R Square	Std. Error of the Estimate
1	.911[a]	.830	.730	17.801

a. Predictors: (Constant), Diastolic Blood Pressure (mmHg), Body Mass Index, Waist-Hip ratio, Systolic Blood Pressure of Cases (mmHg)

b. Dependent Variable: HDL

ANOVA[a]

Model		Sum of Squares	df	Mean Square	F	Sig.
1	Regression	2893.946	4	723.486	2.283	.060[b]
	Residual	125161.016	395	316.863		
	Total	128054.961	399			

a. Dependent Variable: HDL

b. Predictors: (Constant), Diastolic Blood Pressure (mmHg), Body Mass Index, Waist-Hip ratio, Systolic Blood Pressure of Cases (mmHg)

Coefficients[a]

Model		Unstandardized Coefficients		Standardized Coefficients		
		B	Std. Error	Beta	t	Sig.
1	(Constant)	12.935	14.119		.916	.360
	Body Mass Index	-.070	1.101	-.003	-.064	.949
	Waist-Hip ratio	27.963	14.237	-.099	-2.001	.049
	Systolic Blood Pressure of Cases (mmHg)	.121	.056	.140	2.152	.032
	Diastolic Blood Pressure (mmHg)	-.057	.074	-.051	-.771	.441

a. Dependent Variable: HDL

The model in this example is HDL= 0.121(Systolic blood pressure) +27.963(Waist-Hip Ratio) + 12.935. The other two variables are discarded as their *p* values are >0.05. The ANOVA value suggests that our model may not be a good fit to the data we derived it from. The adjusted R-squared (0.73) suggests that our variables together may predict up to 73% of the variability of HDL cholesterol.

The standardized regression coefficients (*β*) allow direct comparison of the relative contribution each predictor variable makes to the outcome. Systolic blood pressure with a *β* of 0.140 makes the greatest contribution while body mass index has the least contribution.

Categorical variables can be included in regression model. A binary variable will be automatically recoded into a *dummy variable* with values of 0 and 1. A categorical variable with 3 or more values will be recoded like multiple response data into several dummy variables before inclusion in the model.

The number of dummy variables created automatically from a categorical variable is equal to the number of possible values minus one. For example, "**Marital status**" with 4 possible values ("**Married**", "**Single**", "**Divorced**" and "**Widowed**") becomes 3 dummy variables in the model (each coded with values: 0, 1). The reason for excluding one of the dummy variables is that its value can always be deduced from the remaining dummy variables.

The table below shows how dummy variables created automatically would look if they actually appeared in a data editor. (We assume "**Widowed**" is the dummy variable excluded from the model. In reality, it could be any of the four. Note that if a different dummy variable is excluded, the model obtained will be different).

	Categorical Variable	Dummy (1)	Dummy (2)	Dummy (3)	*Excluded Dummy*
Case	**MarStat**	**Married**	**Single**	**Divorced**	***Widowed***
1	Single	0	1	0	*0*
2	Married	1	0	0	*0*
3	Divorced	0	0	1	*0*
4	Widowed	0	0	0	*1*

A table similar to this appears as part of the output of the analysis.

Categorical Variables Codings

		Frequency	Parameter coding		
			(1)	(2)	(3)
Marital Status	Divorced	16	1.000	.000	.000
	Married	273	.000	1.000	.000
	Single	45	.000	.000	1.000
	Widowed	66	.000	.000	.000

BINARY LOGISTIC REGRESSION

In these models, the outcome variable is binary (yes/no, dead/alive, male/female, etc). In simple logistic regression, a single predictor variable is used to predict the probability of an outcome. This is also the aim of an odds ratio obtained from a 2 x 2 table ("crude or unadjusted odds ratio"). Unlike crude odds ratio, continuous independent variables are permitted here. In multiple logistic regression, more than one predictor variable is involved.

An example of a simple logistic regression model is this: "How does increasing HbA1c affect the probability of having diabetic nephropathy?"

To carry out this analysis, go to "**Analyze**" >> "**Regression**" >> "**Binary Logistic...**"

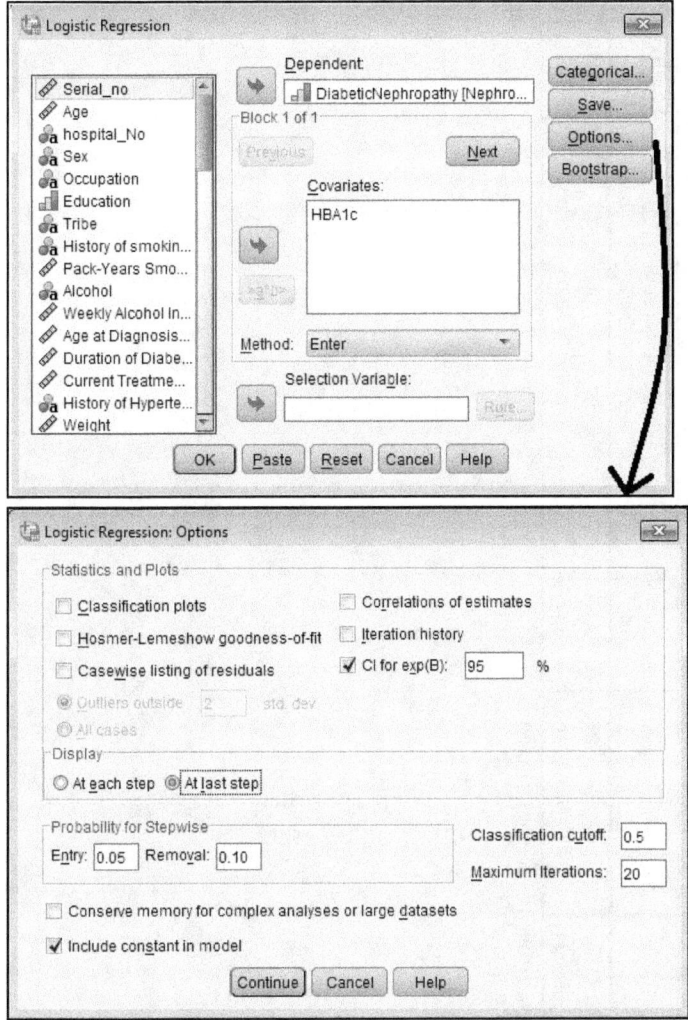

The output in Viewer Window shows several tables.

Variables in the Equation

		B	S.E.	Wald	df	Sig.	Exp(B)	95% C.I.for EXP(B)	
								Lower	Upper
Step 1[a]	HBA1c	-.060	.050	1.415	1	.234	.942	.853	1.040
	Constant	.562	.409	1.886	1	.170	1.754		

a. Variable(s) entered on step 1: HBA1c.

Exp(B) is the *adjusted* odds ratio. In this example, it is not significant (OR=0.942, CI=0.853, 1.040; p=0.234).

For multiple logistic regression, go through the same steps as in simple logistic regression. However, enter the all the desired independent variables into the "**Covariates**" field.

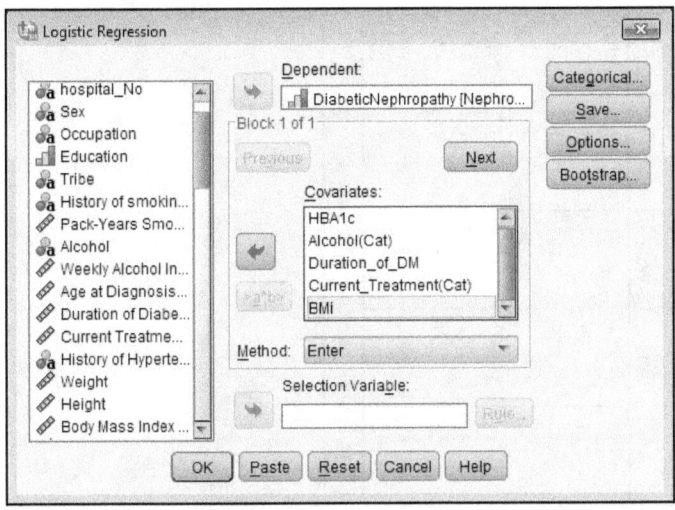

Note that "**Alcohol**" and "**Current_Treatment**" are recognized as categorical variables and automatically recoded for the model.

The results show that BMI and "Use of insulin alone" (Current_Treatment[2]) were significant enough to be included in the model for predicting diabetic nephropathy.

Variables in the Equation

	B	S.E.	Wald	df	Sig.	Exp(B)	95% C.I.for EXP(B) Lower	Upper
Step 1[a] HBA1c	-.075	.054	1.905	1	.167	.928	.834	1.032
Alcohol(1)	.151	.269	.314	1	.575	1.162	.686	1.969
Duration_of_DM	.017	.018	.923	1	.337	1.017	.982	1.054
Current_Treatment			5.395	3	.145			
Current_Treatment(1)	.170	.582	.085	1	.770	1.185	.379	3.708
Current_Treatment(2)	-1.241	.587	4.466	1	.035	.289	.091	.914
Current_Treatment(3)	.239	.333	.518	1	.472	1.270	.662	2.439
BMi	-.130	.025	27.881	1	.000	.878	.836	.921
Constant	4.170	.907	21.147	1	.000	64.720		

a. Variable(s) entered on step 1: HBA1c, Alcohol, Duration_of_DM, Current_Treatment, BMi.

TESTS OF AGREEMENT

These special tests are used to compare multiple observations or measurements of a particular variable carried out on the same sample.

TESTS OF DIAGNOSTIC ACCURACY

A possible scenario is when a new test to detect a condition is developed. One may want to assess how well this new test performs compared to a gold standard test.

Let's assume that a company has developed a test called SpitIt which they claim can detect stomach cancer within minutes using only saliva. The gold standard for diagnosing stomach cancer is histology which requires a hospital procedure and takes time to produce results. SpitIt would be a faster and less distressing method of diagnosing stomach cancer.

A study involving 400 subjects suspected to have stomach cancer was carried out. Saliva was tested with SpitIt. At the same time, histology was performed on the same subjects.

Results from TestIt and histology were obtained and entered into SPSS as variables "**TestIt**" and "**Disease**". A positive TestIt result is interpreted as "**Test Positive**" and recorded as "1" under the variable "**SpitIt**" while a negative result is given a value of "**2**" and labelled as "**Test Negative**".

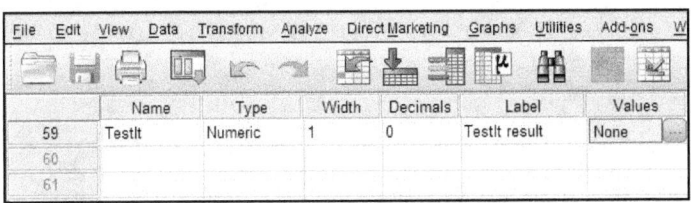

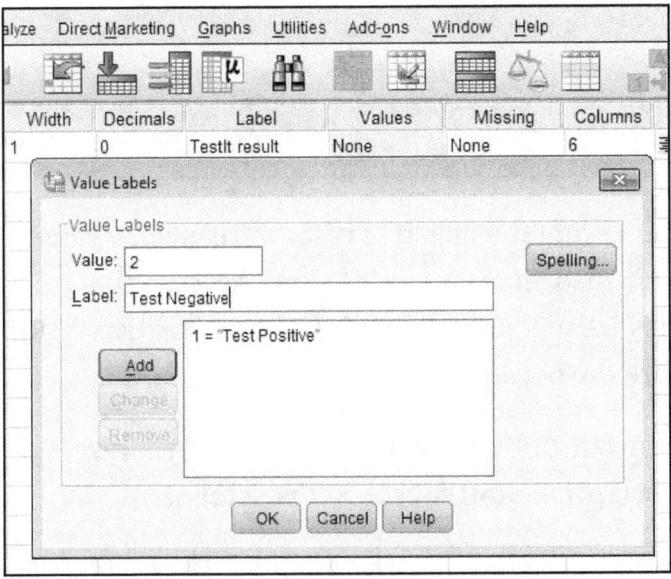

A positive histology result is recorded as "**1**" and labelled as "**Disease Present**" while a negative histology result is denoted as "**Disease Absent**" and given a value of "**2**".

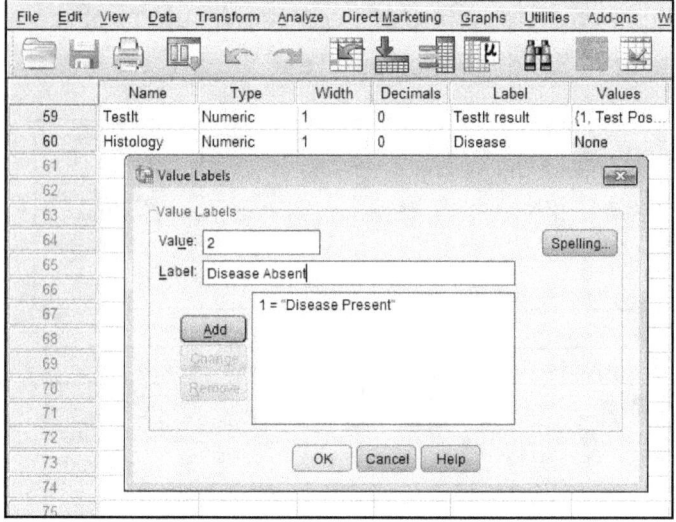

Observe that the negative results are given a value of "**2**" instead of "**0**". This is done so that the cross-tables we will produce will show the positive results before the negative results.

To produce this table, go to "**Analyze**" >> "**Descriptive Statistics**" >> "**Crosstabs...**"

"**TestIt**" occupies the rows while "**Disease**" forms the column variable.

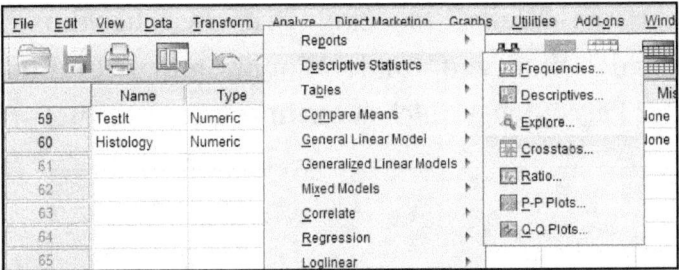

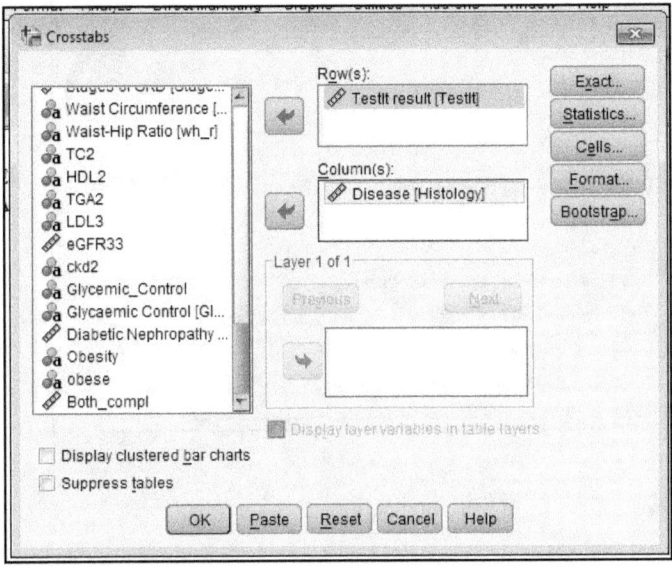

Select the **"Cells"** tab and tick the checkboxes labelled **"Observed"** under <u>Counts</u> and **"Column"** under <u>Percentages</u>. Click "**Continue**".

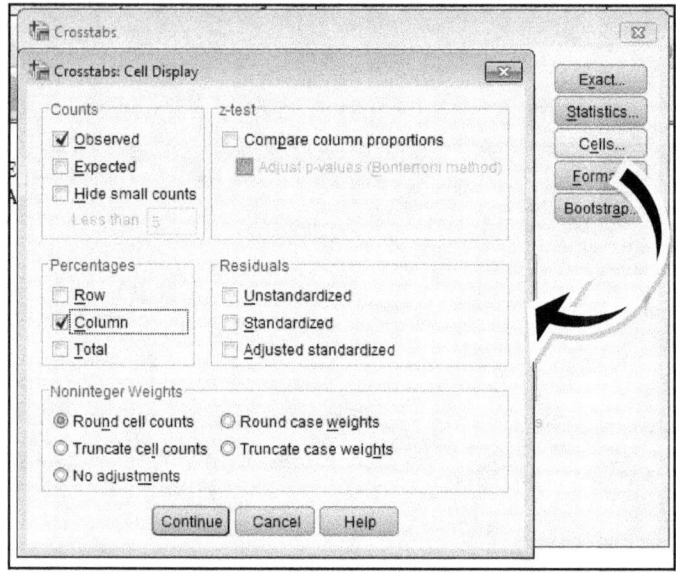

Select the "**Format**" tab, highlight the radio button labelled "**Ascending**" under <u>Row Order</u> and click "**Continue**".

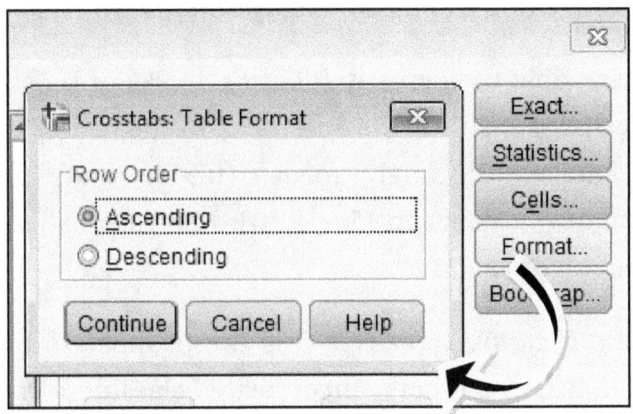

The result of the analysis is shown below:

TestIt result * Disease Crosstabulation

			Disease		
			Disease Present	Disease Absent	Total
TestIt result	Test Positive	Count	127	6	133
		% within Disease	60.8%	3.1%	33.3%
	Test Negative	Count	82	185	267
		% within Disease	39.2%	96.9%	66.8%
Total		Count	209	191	400
		% within Disease	100.0%	100.0%	100.0%

TestIt was able to correctly identify only 60.8% of those who had the disease. This percentage (expressed as a decimal: 0.61) is referred to as the *Sensitivity* of TestIt. It is also called the *True Positive*

Rate (TPR) or *Hit Rate* of the test.

Note that six (3.1%) of the subjects did not have the disease but had positive results with TestIt. This percentage, expressed as a proportion (0.03), is the *False Positive Rate (FPR)* or *False Alarm Rate*.

TestIt was able to correctly identify 96.9% of those *without* disease. This is the *Specificity* of the test. It is expressed in decimal form (0.97) and is mathematically equal to (1 − FPR). It is also called the *True Negative Rate.*

The *False Negative Rate (FNR)* is the proportion of actual disease patients incorrectly classified as negative.

The False Negative Rate is also called the *Miss Rate* of the test. It is equal to (1-Sensitivity).

The *(Positive) Likelihood Ratio* is the ratio of the true positive rate to the false positive rate. It is obtained by calculating:

$$\frac{Sensitivity}{False\ Positive\ Rate}$$

Our example gives us: 0.61/0.03 ≈ 20.

A *Negative Likelihood Ratio* is calculated by dividing

the

$$\frac{False\ Negative\ Rate}{Specificity}$$

Likelihood ratios can be used to compare the relative usefulness of different tests. The higher the positive likelihood ratio, the more useful the test. Accuracy of a test is the number of correctly identified cases divided by the total population. In our example, True Negative + True Positive divided by 400 = (127 +185)/400 = 0.78.

To calculate the *Predictive Values,* create the cross-tables the same way we did for specificity and sensitivity analysis. However, after clicking the "**Cells**" tab, select "**Row**" under "Percentages".

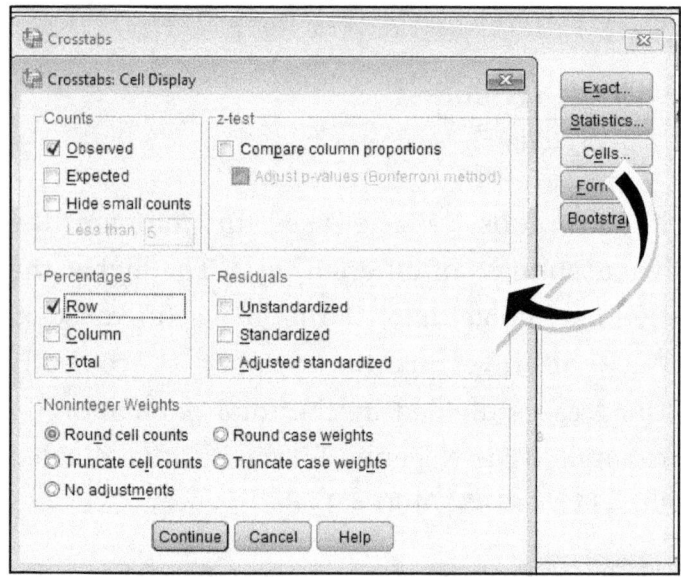

The resulting cross-table showing row percentages is shown below:

Testit result * Disease Crosstabulation

			Disease		
			Disease Present	Disease Absent	Total
Testit result	Test Positive	Count	127	6	133
		% within Testit result	95.5%	4.5%	100.0%
	Test Negative	Count	82	185	267
		% within Testit result	30.7%	69.3%	100.0%
Total		Count	209	191	400
		% within Testit result	52.3%	47.8%	100.0%

The *Positive Predictive Value (PPV)* is the percentage of subjects who test positive that actually have the disease. In the example above, 133 subjects had a positive test result but only 95.5% of them actually had the disease. The PPV is therefore 0.96 (95.5%).

The *Negative Predictive Value (NPV)* is the percentage of subjects who test negative that do not have the disease. Of the 267 test-negative subjects in our example, only 69.3% of them were actually disease-free. The NPV in this example is 0.69.

The drawback of predictive values is that they cannot be used to directly compare different tests. Sensitivity, specificity and likelihood ratio permit direct comparison of multiple tests.

The 2 x 2 contingency table below is called the *confusion matrix.* It has been expanded to show the relevant statistics that can be calculated and their formulas.

		Condition (as determined by "Gold standard")		
Test outcome	Total population	Condition positive	Condition negative	Prevalence = $\dfrac{\Sigma\ \text{Condition positive}}{\Sigma\ \text{Total population}}$
Test outcome positive	True positive	False positive (Type I error)	Positive predictive value (PPV), Precision = $\dfrac{\Sigma\ \text{True positive}}{\Sigma\ \text{Test outcome positive}}$	False discovery rate (FDR) = $\dfrac{\Sigma\ \text{False positive}}{\Sigma\ \text{Test outcome positive}}$
Test outcome negative	False negative (Type II error)	True negative	False omission rate (FOR) = $\dfrac{\Sigma\ \text{False negative}}{\Sigma\ \text{Test outcome negative}}$	Negative predictive value (NPV) = $\dfrac{\Sigma\ \text{True negative}}{\Sigma\ \text{Test outcome negative}}$
Accuracy (ACC) = $\dfrac{\Sigma\ \text{True positive} + \Sigma\ \text{True negative}}{\Sigma\ \text{Total population}}$	True positive rate (TPR), Sensitivity, Recall = $\dfrac{\Sigma\ \text{True positive}}{\Sigma\ \text{Condition positive}}$	False positive rate (FPR), Fall-out = $\dfrac{\Sigma\ \text{False positive}}{\Sigma\ \text{Condition negative}}$	Positive likelihood ratio (LR+) = $\dfrac{\text{TPR}}{\text{FPR}}$	Diagnostic odds ratio (DOR) = $\dfrac{\text{LR+}}{\text{LR-}}$
	False negative rate (FNR), Miss rate = $\dfrac{\Sigma\ \text{False negative}}{\Sigma\ \text{Condition positive}}$	True negative rate (TNR), Specificity (SPC) = $\dfrac{\Sigma\ \text{True negative}}{\Sigma\ \text{Condition negative}}$	Negative likelihood ratio (LR-) = $\dfrac{\text{FNR}}{\text{TNR}}$	

RECEIVER OPERATING CHARACTERISTICS (ROC) CURVES

The example above involved a test with only two possible values (positive and negative).

Sensitivity and specificity can also be calculated for a test that has continuous values. This is done by comparing the test's performance at multiple cut-off values against the results from the gold standard.

The ROC Curve is a plot of true positive rate of a diagnostic test against its false positive rate at different cut-off points (*thresholds*). This is the same as plotting *Sensitivity* versus 1 – *Specificity.*

ROC curves can be used to compare diagnostic tests. They can also be used to determine an optimal *cut-off value* between normal and abnormal results for a particular test.

Different diagnostic tests can be compared by plotting their ROC curves on the same axes. The closer an ROC curve is to the upper left corner of the graph, the better it is. Such a curve will have a greater Area Under the Curve (AUC). AUC is calculated by SPSS and is expressed as a decimal fraction (less than 1).

Scenario: A biomedical company has found that an enzyme found in saliva called contase can be used to diagnose stomach cancer. Patients with stomach cancer have higher levels of contuse than normal persons but no cut-off value has been established. Levels of contase for 400 subjects have been collected into the variable "**TestCont**".

To carry out ROC analysis, go to "**Analysis**" >> "**ROC...**" In the dialog window that appears, set TestCont as the Test Variable and the definitive result (Histology) as the State Variable. Positive histology results have been coded as "**1**" so the Value of the State Variable is set as "**1**".

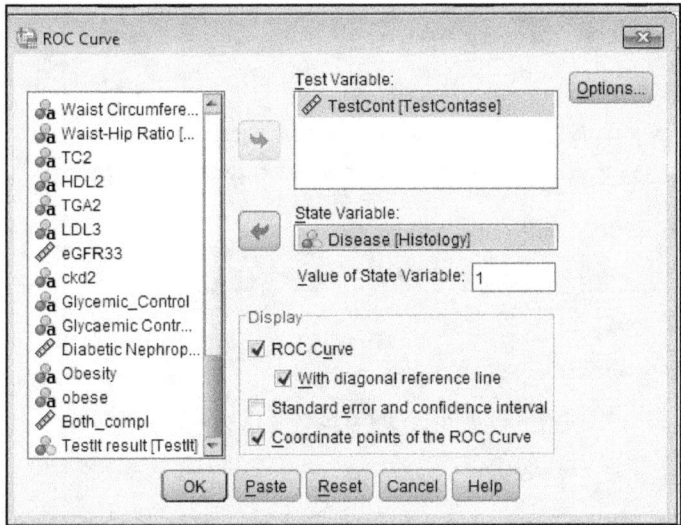

Tick the checkboxes: ROC Curve, With Diagonal Reference Line, and Coordinate Points of the ROC Curve shown on the <u>Display</u> panel.

Click the "**Options...**" tab and select the radio button marked "**Smaller test result indicates more positive result**". Click the tab "**Continue**". This returns you to the main dialog window.

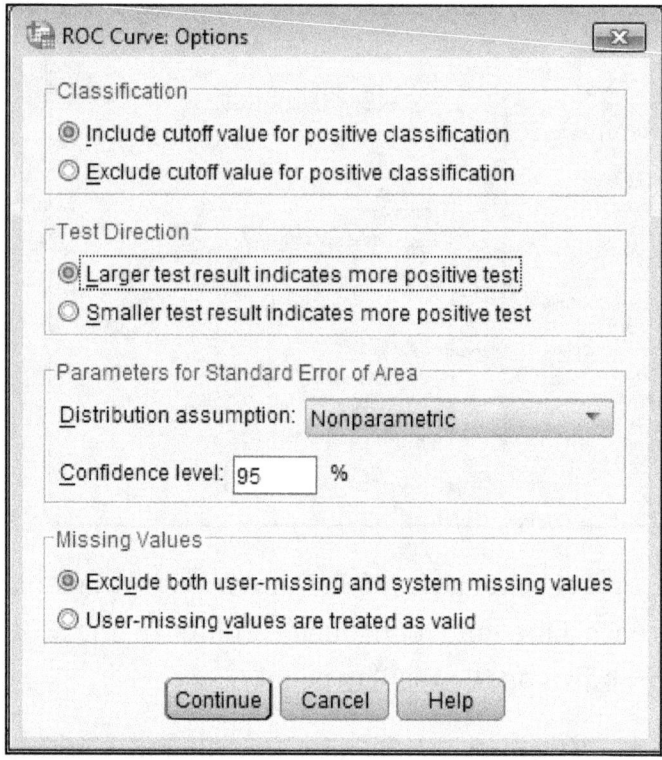

Click the tab marked "**OK**".

The output is shown below:

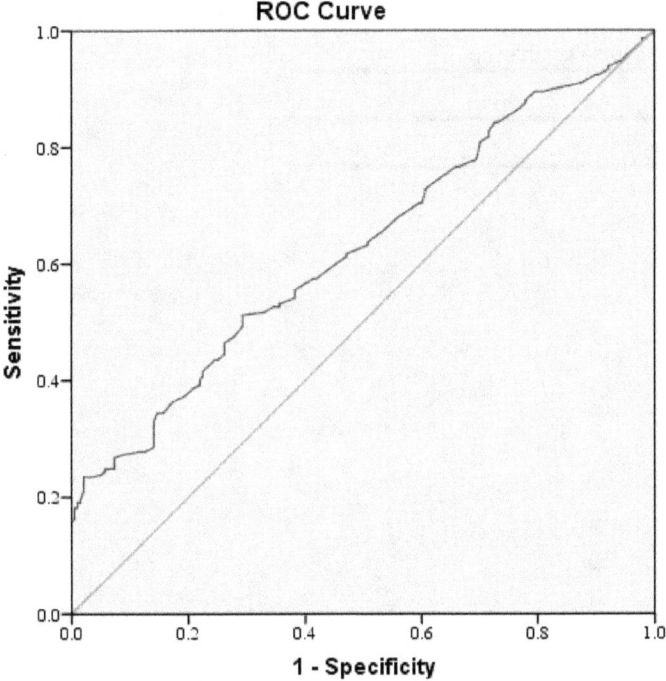

Area Under the Curve

Test Result Variable(s): TestCont

Area
0.627

Coordinates of the Curve

Test Result Variable(s): TestCont

Positive if Greater Than or Equal To[a]	Sensitivity	1 - Specificity
7.9000	1.000	1.000
9.4500	.995	1.000
10.5000	.990	.990
11.2850	.986	.984
12.0150	.986	.979
12.7300	.981	.979
13.5000	.967	.963
14.5000	.952	.948
15.0650	.947	.948
15.5650	.947	.942
134.7500	.062	.000
138.0000	.057	.000
141.5000	.048	.000
143.5000	.038	.000
146.5000	.033	.000
149.5000	.029	.000
165.0000	.019	.000
186.0000	.014	.000
240.0000	.005	.000
289.0000	.000	.000

The first output of the analysis is the ROC curve showing a diagonal reference line. The curve shown here is above and to the left of the reference line. This indicates that TestContase has some usefulness in detecting stomach cancer

An ROC curve that closely follows the reference line will have an Area under the curve of 0.5 (50%). Such a curve would be useless in differentiating positive from negative results as every cut-off value would have an equal chance (50/50) of being either positive or negative.

An ROC that lies below the reference line is said to be *perverse* as it does the opposite of what it is intended to. A perverse ROC curve can be made useful by inverting its values (values *below* the cut-off imply the disease is present).

An example of a perverse ROC is shown below.

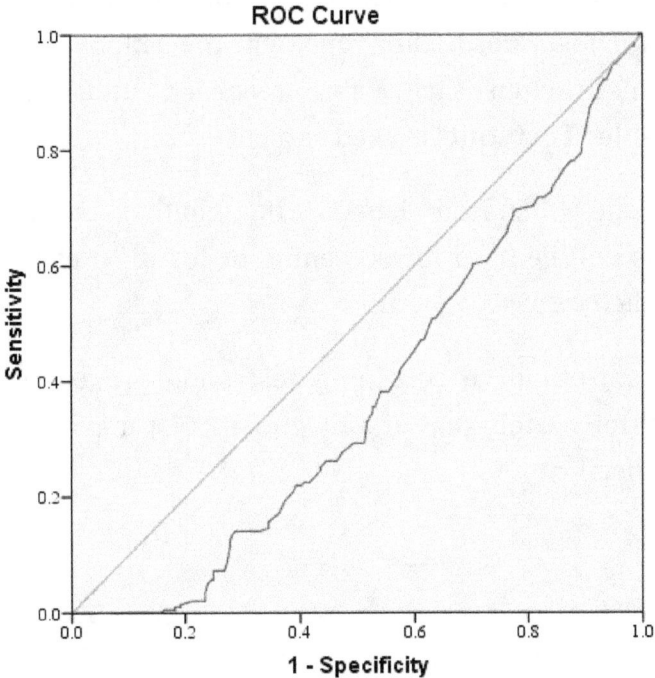

To invert a perverse ROC in SPSS, repeat the analysis. However, this time, under the "**Options...**" tab, select the radio button "**Smaller test result indicates more positive result**".

The Area under the Curve is also displayed in the output. A perfect test has an AUC of 1.

To save space, only the first and last 20 rows of the table "**Coordinates of the Curve**" are shown in this book. These coordinates represent the TPR and FPR obtained when each value recorded under the variable "**TestCont**" is used as a cut-off.

The cut-off chosen depends on whether Contase levels will be used for screening or for diagnosis of stomach cancer.

The cut-off for a screening test should maximise sensitivity while that of a diagnostic test maximises the specificity.

The usefulness of multiple tests for a condition can be compared by plotting their ROC curves on the same axes.

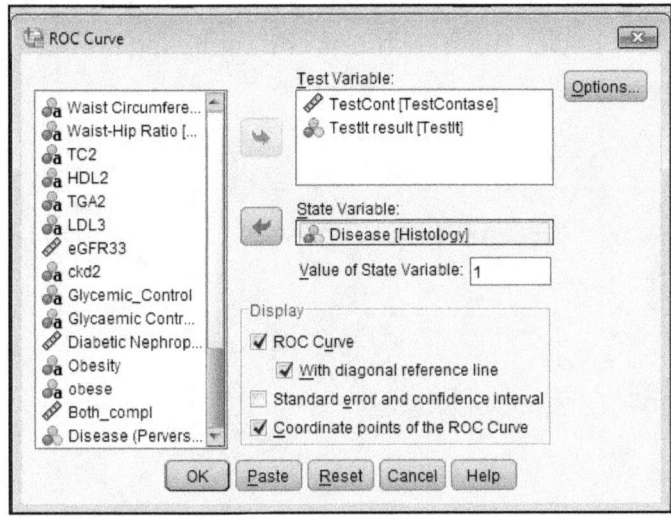

The resulting ROC curves are shown below. TestIt produces a "curve" that is closer to the left upper corner. This translates to a greater AUC (0.788 against 0.627) as can be seen from the output table. TestIt is better than TestCont at diagnosing stomach cancer.

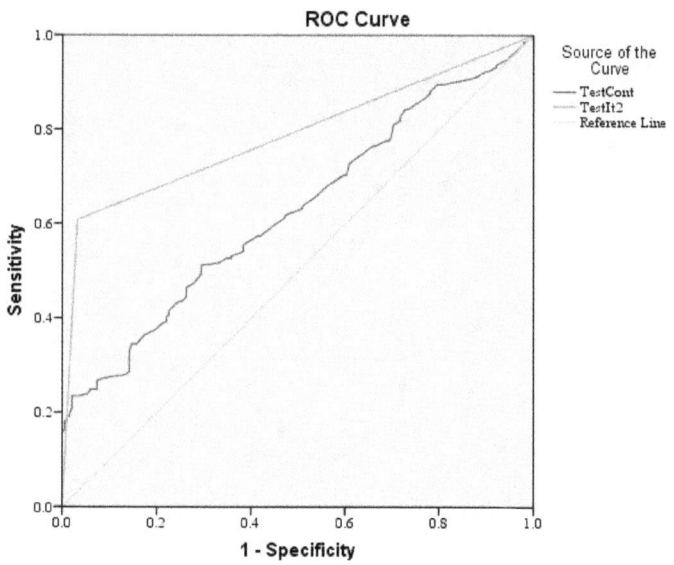

Area Under the Curve

Test Result Variable(s)	Area
TestCont	.627
TestIt	.788

KAPPA

There are situations where different raters independently provide their assessments or diagnoses.

Kappa tests how closely these ratings agree after removing the effect of chance.

Kappa comes in handy when there is no established gold standard for making a diagnosis.

Two expert physicians may be asked to diagnose 400 patients to determine those that have stomach cancer and those that don't.

For each subject, each rater makes a diagnosis of "**Yes**" or "**No**" coded as "**0**" and "**1**" respectively. The responses for each rater occupy a column (variable "**Doctor A**", "**DoctorB**").

To calculate Kappa, go to "**Analyze**" >> "**Descriptive Statistics**" >> "**Crosstabs...**"

In the dialog window that appears, send any of the two variables to the "**Row(s)**" field and the other to the "**Column(s)**" field.

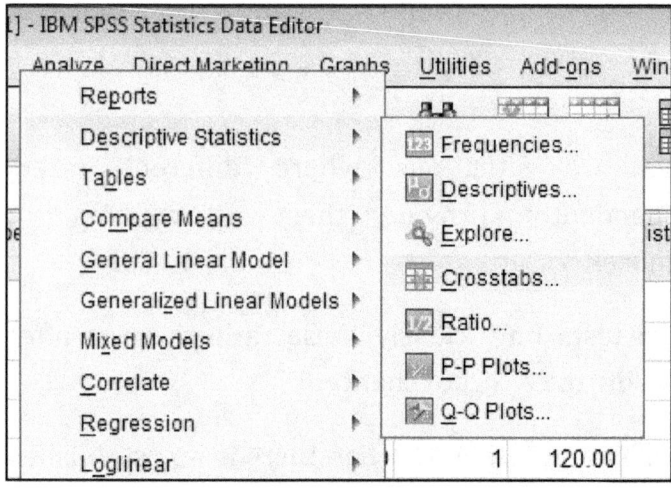

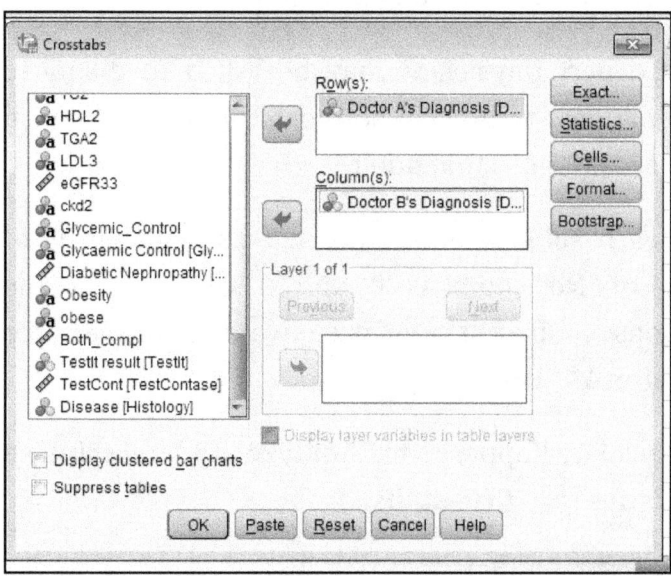

Click the "**Statistics...**" button. In the dialog box that appears, select the checkbox labelled "**Kappa**". Click "**Continue**" to close the dialog box.

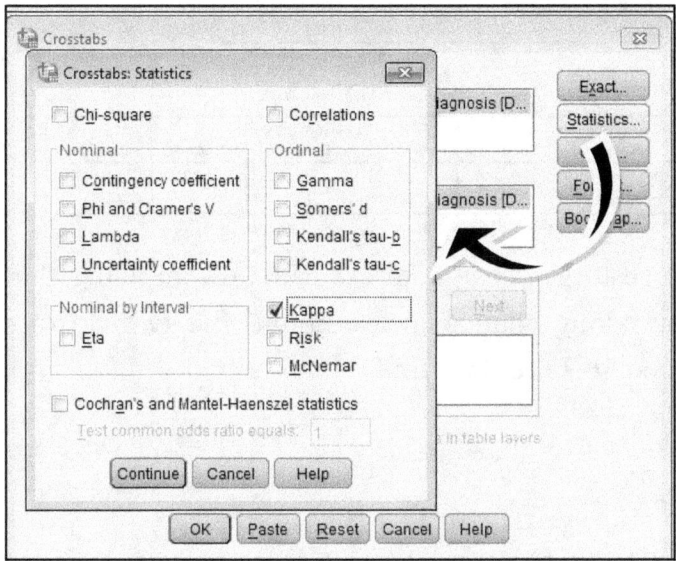

Click "**OK**" in the main dialog window to run the analysis.

The results of the analysis are appear in the output file and are shown below.

Doctor A's Diagnosis * Doctor B's Diagnosis Crosstabulation

Count

		Doctor B's Diagnosis		Total
		No	Yes	
Doctor A's Diagnosis	No	185	82	267
	Yes	6	127	133
Total		191	209	400

Symmetric Measures

		Value	Asymp. Std. Error[a]	Approx. T[b]	Approx. Sig.
Measure of Agreement	Kappa	.567	.038	12.219	.000
N of Valid Cases		400			

The results reveal a kappa of 0.567 implying that after taking chance into account, the two doctors agreed 56.7% of the time.

SURVIVAL ANALYSIS

Survival analysis is concerned with the time it takes for an event to occur and the variables associated with it. An event is a change in the value of a variable occurring at a specific point in time.

A few examples of events include death, admission, marriage, divorce, discharge from hospital, commencement of treatment and childbirth.

The term "censored case" is used to describe a subject who is not known to have experienced the event of interest during the study period. This may be because the event did not occur in them during this period. It may also be the result of inadequate data for these subjects due to loss to follow-up.

LIFE TABLES

These are also called *mortality tables* or *actuarial tables* (due to their importance in insurance and actuarial science). Classically, life tables are used to predict the probability of a person of a particular age dying before his or her next birthday.

Life tables divide the period of interest into smaller time intervals and assess the probability of the event happening within each time interval.

Life tables can be created for the entire dataset. Categorization may also be introduced; this permits comparison of the survival characteristics of different subgroups.

A possible scenario is one where we have a cohort of 400 stomach cancer patients being followed up from when they were diagnosed until when they die.

A variable "**Outcome**" was used to encode whether the subject is known to have died during the study period ("**1**") or not ("**0**"). Time from diagnosis to death (in days) was coded as the categorical variable

"TimeTillDeath".

We can create a life table for the entire population looking at survival within 30-day intervals. We may also decide to do a subgroup analysis using Level of Education ("**Education**") as a factor.

To create a life table for the entire cohort, go to "**Analyze**" >> "**Life Tables**" >> "**Life Tables...**"

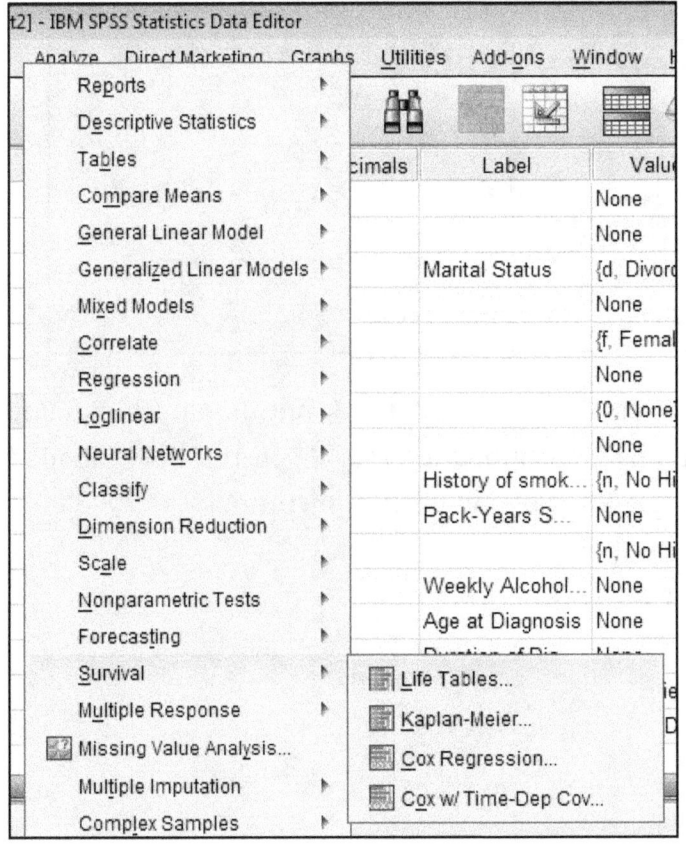

In the dialog window that appears, push the variable **TimeTillDeath** into the field marked "**Time:**"

Select the variable "**Outcome**" into the field marked "**Status:**"

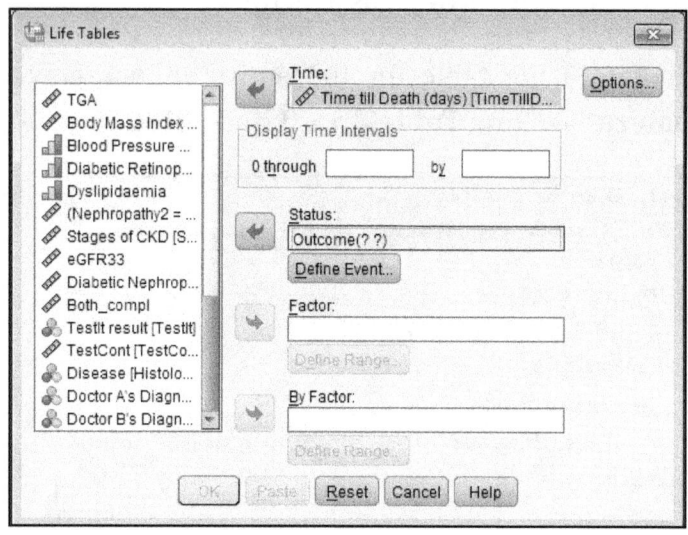

Click the "**Define Event...**" button. In the dialog window that appears, type "**1**" in the field labelled "**Single Value:**" and click "**Continue**".

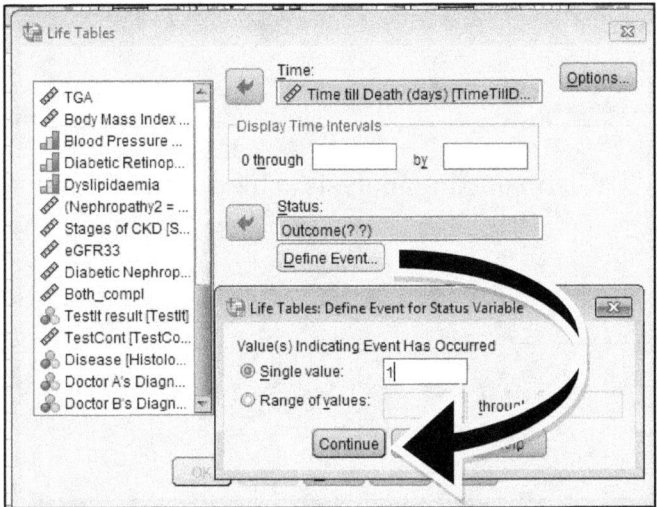

In the main dialog window, there is a group labelled "**Display Time Intervals:**" containing two fields. In the field preceded by the label "**0 through:____**", type in the maximum value for the Time variable. Exploring this dataset shows us that the maximum time till death is 421 days.

In the second field that follows the label "**_by_**", type in the time interval. In this example, we are looking at 30-day intervals, so type in "**30**".

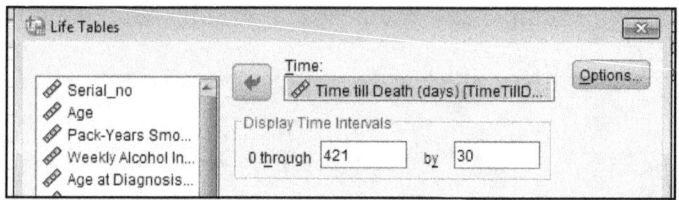

Click "**OK**" to run the analysis. The output is shown below.

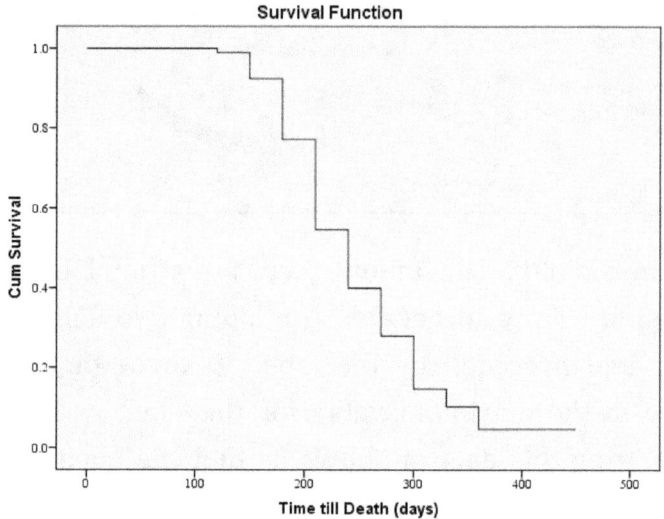

The Survival function curve plots the decrease in cumulative survival proportion as "**Time till Death**" increases.

QUICK & EASY STATISTICS

Life Table[a]

Interval Start Time	Number Entering Interval	Number Withdrawing during Interval	Number Exposed to Risk	Number of Terminal Events	Proportion Terminating	Proportion Surviving	Cumulative Proportion Surviving at End of Interval	Std. Error of Cumulative Proportion Surviving at End of Interval	Probability Density	Std. Error of Probability Density	Hazard Rate	Std. Error of Hazard Rate
0	400	0	400.000	0	0.00	1.00	1.00	0.00	0.000	0.000	0.00	0.00
30	400	0	400.000	0	0.00	1.00	1.00	0.00	0.000	0.000	0.00	0.00
60	400	0	400.000	0	0.00	1.00	1.00	0.00	0.000	0.000	0.00	0.00
90	400	5	397.500	4	.01	.99	.99	.01	.000	.000	.00	.00
120	391	23	379.500	25	.07	.93	.92	.01	.002	.000	.00	.00
150	343	44	321.000	53	.17	.83	.77	.02	.005	.001	.01	.00
180	246	67	212.500	62	.29	.71	.55	.03	.008	.001	.01	.00
210	117	26	104.000	28	.27	.73	.40	.03	.005	.001	.01	.00
240	63	12	57.000	17	.30	.70	.28	.03	.004	.001	.01	.00
270	34	9	29.500	14	.47	.53	.15	.03	.004	.001	.02	.01
300	11	2	10.000	3	.30	.70	.10	.03	.001	.001	.01	.01
330	6	1	5.500	3	.55	.45	.05	.03	.002	.001	.03	.01
360	2	1	1.500	0	0.00	1.00	.05	.03	0.000	0.000	0.00	0.00
390	1	0	1.000	0	0.00	1.00	.05	.03	0.000	0.000	0.00	0.00
420	1	1	.500	0	0.00	1.00	.05	.03	0.000	0.000	0.00	0.00

a. The median survival time is 219.53

167

From the life tables, we can tell that most of the deaths occur between 150 and 180 days after diagnosis (highest *number of terminal events*). These intervals also have the highest *probability densities* (probability of experiencing the event during a particular interval).

Median survival period was 219.5 days (half the patients survived to 219 days).

We can create a life table that investigates the role of a factor (in this scenario, level of education). To do this, repeat all the steps above except the final click of the "**OK**" button.

Push the variable "**Education**" into the field labelled "**Factor:**"

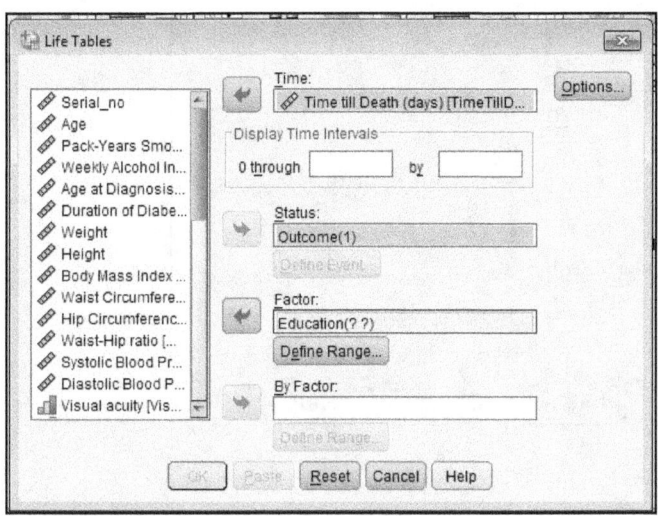

Click the "**Options...**" button. A dialog box appears in which the checkbox "**Life Tables**" is ticked by default. Tick the checkboxes for the desired plots ("**Survival**", etc). Select the radio button "**Pairwise**" under the group "**Compare Levels of First Factor**". Click **Continue** to return to the main dialog window.

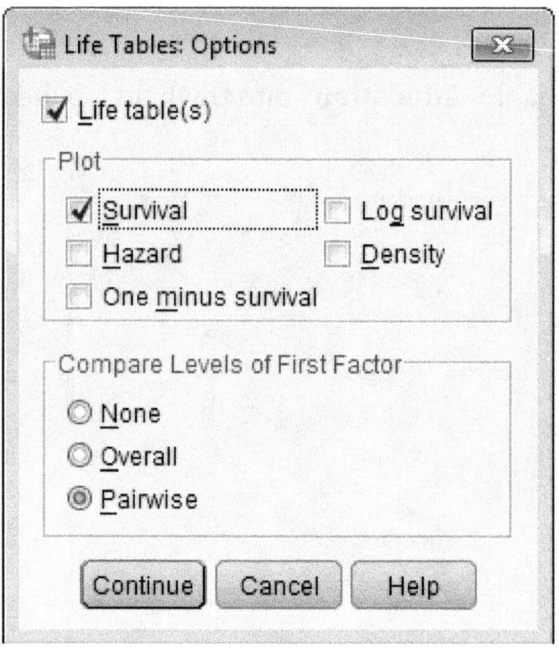

Click the button labelled "**Define Range...**" Type in "**0**" in the field labelled "**Minimum:**" and "**3**" in the field marked "**Maximum**". (The range of values for the variable "**Education**" in this dataset is 0 through 3). Click "**Continue**" to return to the main dialog window.

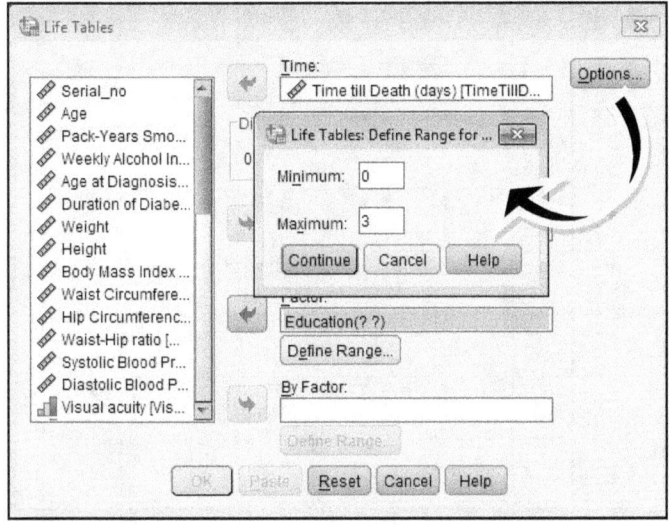

Click "**OK**" to perform the analysis.

The resulting life table is shown below (truncated to fit the page).

Life Table

First-order Controls / Education		Number Entering Interval	Number Withdrawing during Interval	Number Exposed to Risk	Number of Terminal Events	Proportion Terminating	Proportion Surviving	Cumulative Proportion Surviving at End of Interval	Std. Error of Cumulative Proportion Surviving at End of Interval	Probability Density	Std. Error of Probability Density	Hazard Rate	Std. Error of Hazard Rate
None	0	103	0	103.000	0	0.00	1.00	1.00	0.00	0.000	0.000	0.00	0.00
	30	103	0	103.000	0	0.00	1.00	1.00	0.00	0.000	0.000	0.00	0.00
	60	103	0	103.000	0	0.00	1.00	1.00	0.00	0.000	0.000	0.00	0.00
	240	16	4	14.000	4	.29	.71	.25	.06	.003	.001	.01	.01
	270	8	1	7.500	5	.67	.33	.08	.05	.006	.002	.03	.01
	300	2	0	2.000	1	.50	.50	.04	.04	.001	.001	.02	.02
	330	1	0	1.000	1	1.00	0.00	0.00	0.00	.001	.001	.07	.00
Primary	0	108	0	108.000	0	0.00	1.00	1.00	0.00	0.000	0.000	0.00	0.00
	30	108	0	108.000	0	0.00	1.00	1.00	0.00	0.000	0.000	0.00	0.00
	60	108	0	108.000	0	0.00	1.00	1.00	0.00	0.000	0.000	0.00	0.00
	330	3	0	3.000	1	.33	.67	.10	.05	.002	.001	.01	.01
	360	2	1	1.500	0	0.00	1.00	.10	.05	0.000	0.000	0.00	0.00
	390	1	0	1.000	0	0.00	1.00	.10	.05	0.000	0.000	0.00	0.00
	420	1	1	.500	0	0.00	1.00	.10	.05	0.000	0.000	0.00	0.00
Secondary	0	112	0	112.000	0	0.00	1.00	1.00	0.00	0.000	0.000	0.00	0.00
	30	112	0	112.000	0	0.00	1.00	1.00	0.00	0.000	0.000	0.00	0.00
	60	112	0	112.000	0	0.00	1.00	1.00	0.00	0.000	0.000	0.00	0.00
	270	10	2	9.000	4	.44	.56	.14	.05	.004	.002	.02	.01
	300	4	0	4.000	2	.50	.50	.07	.04	.002	.001	.02	.01
	330	2	1	1.500	1	.67	.33	.02	.03	.002	.001	.03	.03
Tertiary	0	77	0	77.000	0	0.00	1.00	1.00	0.00	0.000	0.000	0.00	0.00
	30	77	0	77.000	0	0.00	1.00	1.00	0.00	0.000	0.000	0.00	0.00
	60	77	0	77.000	0	0.00	1.00	1.00	0.00	0.000	0.000	0.00	0.00
	90	77	4	75.000	0	0.00	1.00	1.00	0.00	0.000	0.000	0.00	0.00
	240	13	2	12.000	3	.25	.75	.52	.10	.006	.003	.02	.01
	270	8	5	5.500	2	.36	.64	.33	.13	.006	.004	.03	.01
	300	1	1	.500	0	0.00	1.00	.33	.13	0.000	0.000	0.00	0.00

Median Survival Time

First-order Controls		Med Time
Education	None	201.26
	Primary	218.08
	Secondary	221.56
	Tertiary	273.33

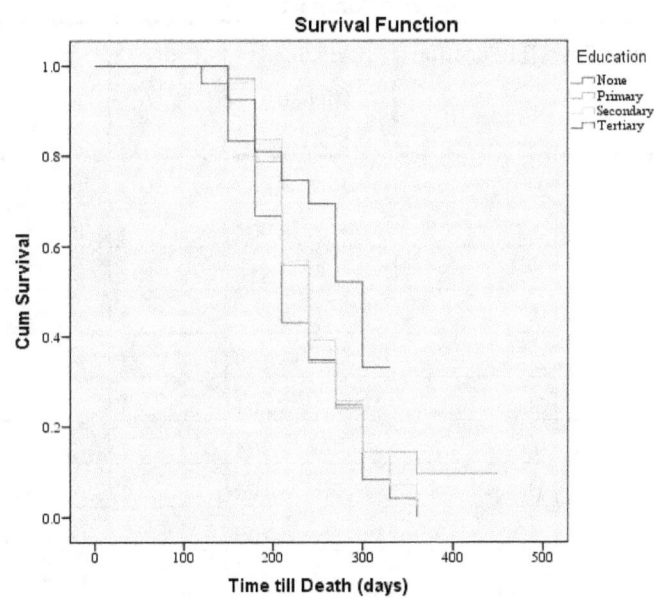

Overall Comparisons[a]

Wilcoxon (Gehan) Statistic	df	Sig.
15.430	3	.001

a. Comparisons are exact.

Pairwise Comparisons[a]

(I) Education	(J) Education	Wilcoxon (Gehan) Statistic	df	Sig.
0	1	6.687	1	.010
	2	8.904	1	.003
	3	9.510	1	.002
1	0	6.687	1	.010
	2	.189	1	.664
	3	1.201	1	.273
2	0	8.904	1	.003
	1	.189	1	.664
	3	.709	1	.400
3	0	9.510	1	.002
	1	1.201	1	.273
	2	.709	1	.400

a. Comparisons are exact.

The results show that subjects with tertiary education had the longest median survival time (273.3 days).

The Wilcoxon (Gehan) statistic is significant implying that there are significant differences in survival times among subjects with different levels of education.

The pairwise comparisons tell us which pairs of groups significantly differ. Those with no education ("**0**") differ significantly from subjects of each of the three groups. Other significant comparisons on the table have been given a dark background by the author, not by SPSS.

KAPLAN-MEIER ANALYSIS

Unlike life tables, Kaplan-Meier analysis does not divide the time-to-event into discrete intervals. Instead the exact times of the event are used to calculate cumulative survival and to compare factors or subgroups.

Kaplan-Meier analysis also takes censored cases into consideration.

Using the earlier scenario, we can perform a Kaplan-Meier analysis because we have the exact time to death of the uncensored subjects.

To perform the procedure on the entire cohort, go to "**Analyze**" >> "**Survival**" >> "**Kaplan-Meier...**"

Select "**TimeTillDeath**" as the "**Time:**" variable and "**Outcome:**" as the "**Status:**" variable.

Click the button labelled "**Define Event...**" and type "**1**" in the field marked "**Single Value:**" in the dialog box that appears. Click "**Continue**" to return to the main dialog window. Click the "**Options...**" button and tick the checkboxes for desired statistics and plots. Click "**Continue**" to return to the main window and then click "**OK**" to produce the output.

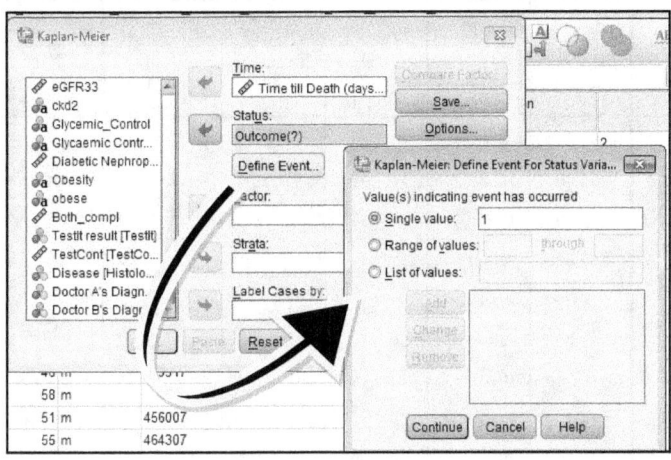

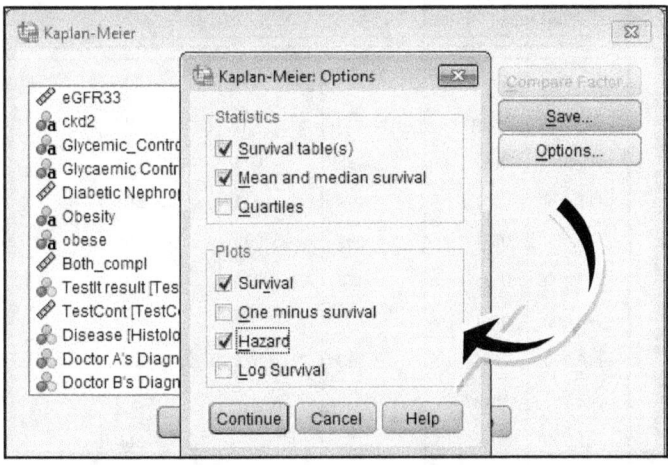

Survival Table

	Time	Status	Cumulative Proportion Surviving at the Time		N of Cumulative Events	N of Remaining Cases
			Estimate	Std. Error		
1	91.000	Alive?			0	399
2	95.800	Died	.997	.003	1	398
3	105.000	Died	.995	.004	2	397
4	107.000	Alive?			2	396
5	112.000	Alive?			2	395
6	113.000	Alive?			2	394
7	114.300	Died	.992	.004	3	393
8	114.300	Alive?			3	392
9	119.700	Died	.990	.005	4	391
10	122.800	Alive?			4	390
390	304.000	Alive?			203	10
391	309.000	Alive?			203	9
392	310.500	Died	.115	.031	204	8
393	311.400	Died	.101	.031	205	7
394	322.000	Died	.087	.029	206	6
395	345.000	Died	.072	.028	207	5
396	346.000	Died			208	4
397	346.000	Died	.043	.023	209	3
398	348.000	Alive?			209	2
399	381.000	Alive?			209	1
400	421.000	Alive?			209	0

Means and Medians for Survival Time

Mean[a]				Median			
Estimate	Std. Error	95% Confidence Interval		Estimate	Std. Error	95% Confidence Interval	
		Lower Bound	Upper Bound			Lower Bound	Upper Bound
231.921	4.873	222.370	241.472	218.000	6.394	205.467	230.533

a. Estimation is limited to the largest survival time if it is censored.

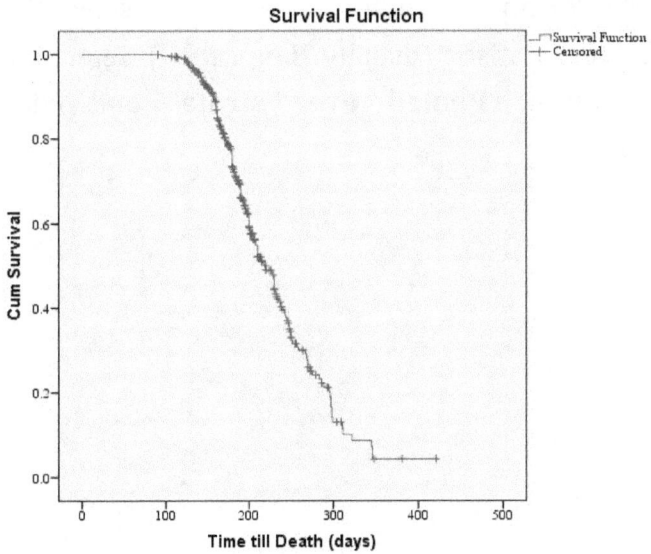

Survival Function

The results show that the median survival time is 218 days using this procedure (similar to the 219 days from the life table).

To compare survival across groups (using "**Education**" as a factor), repeat the procedure shown previously. This time, select "**Education**" as the "**Factor:**" variable.

Click the "**Compare Factor...**" button and select the desired test statistic (usually "**Log rank**"). Select the radio button "**Pooled over strata**" and click "**Continue**".

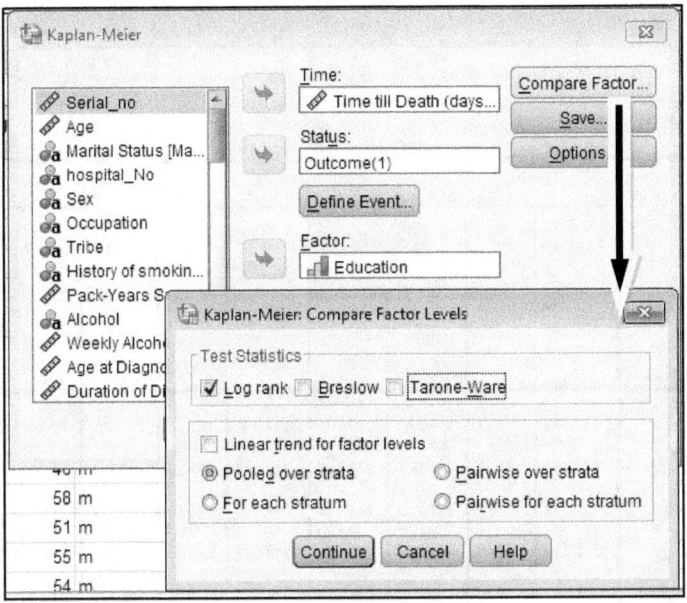

Performing the analysis gives us a survival table with subcategories for each level of education (not shown here), the survival curves, median and means for each group and the log rank overall comparison.

Means and Medians for Survival Time

Education	Mean[a]				Median			
	Estimate	Std. Error	95% Confidence Interval		Estimate	Std. Error	95% Confidence Interval	
			Lower Bound	Upper Bound			Lower Bound	Upper Bound
None	213.124	7.133	199.144	227.104	196.000	7.353	181.589	210.411
Primary	235.951	10.135	216.086	255.817	210.000	5.754	198.721	221.279
Secondary	231.298	6.727	218.113	244.483	227.000	9.012	209.337	244.663
Tertiary	252.132	9.580	233.355	270.909	267.000	16.334	234.985	299.015
Overall	231.921	4.873	222.370	241.472	218.000	6.394	205.467	230.533

a. Estimation is limited to the largest survival time if it is censored.

Overall Comparisons

	Chi-Square	df	Sig.
Log Rank (Mantel-Cox)	11.263	3	.010

Test of equality of survival distributions for the different levels of Education.

Survival Functions

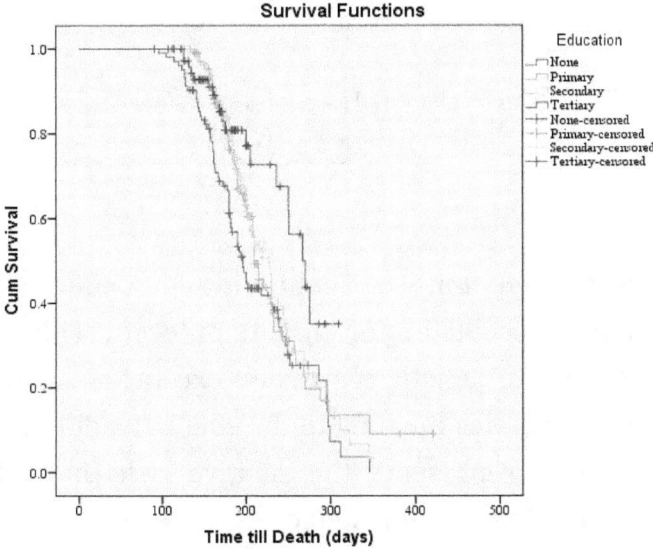

The Log rank test shows that there is a significant difference in time to death among the different levels of education.

COX PROPORTIONAL HAZARDS MODELLING (COX REGRESSION)

Cox regression aims at identifying the effect of independent predictors on time to an event. This is quite similar to logistic regression except that here the model includes the time to the event. In addition, **Exp(B)** here represents the relative risk (hazard ratio) rather than an adjusted odds ratio.

For the cohort of patients with stomach cancer, we can investigate the effects of Marital Status (**"MarStat"**), Education (**"Education"**) and use of a drug (**"Cancidin"**) on time to death.

To perform the Cox regression, go to **"Analyze"** >> **"Survival"** >> **"Cox Regression..."**

Select **"TimeTillDeath"** as the **"Time:"** variable and **"Outcome"** as the **"Status:"** variable. Click the **"Define Event..."**button and type **"1"** in the field labelled **"Single Value:"** Click **"Continue"** to return to the main dialog window.

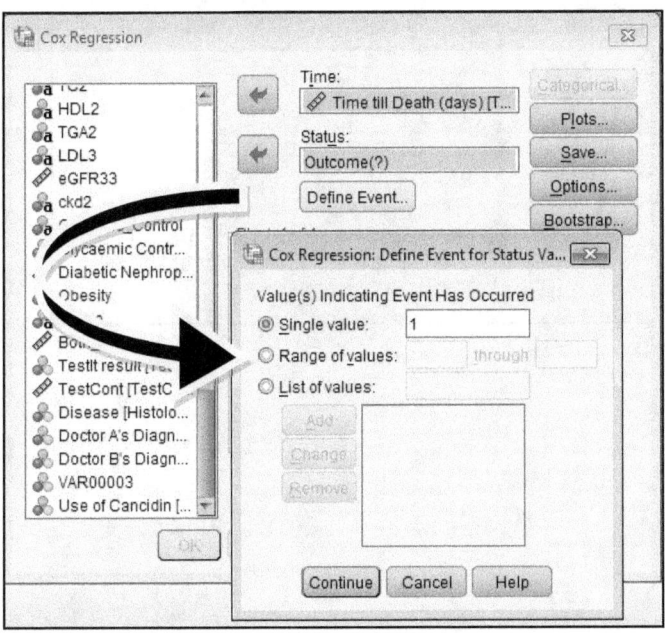

Select **"MarStat"**, **"Education"** and **"Cancidin"** as **"Covariates"**.

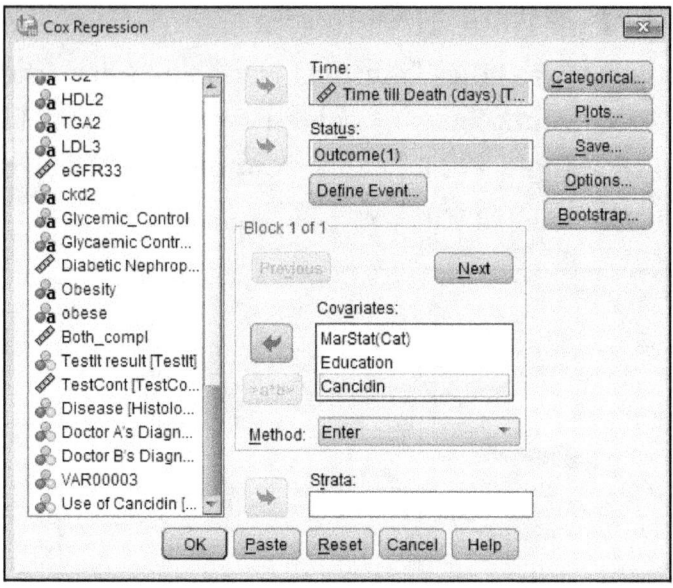

Click the button "**Categorical**" and select "**MarStat**" and "**Education**" into the field "**Categorical Covariates**". Click "**Continue**" to return to the main dialog window.

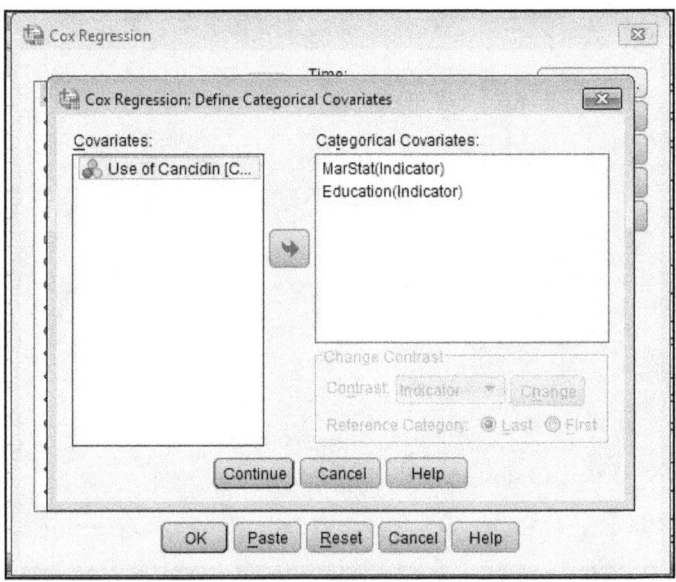

Click the "**Options...**" button and tick the checkbox labelled "**CI for Exp(B)**" to ensure 95% confidence intervals are displayed in the results. Click "**Continue**" to return to the main dialog window.

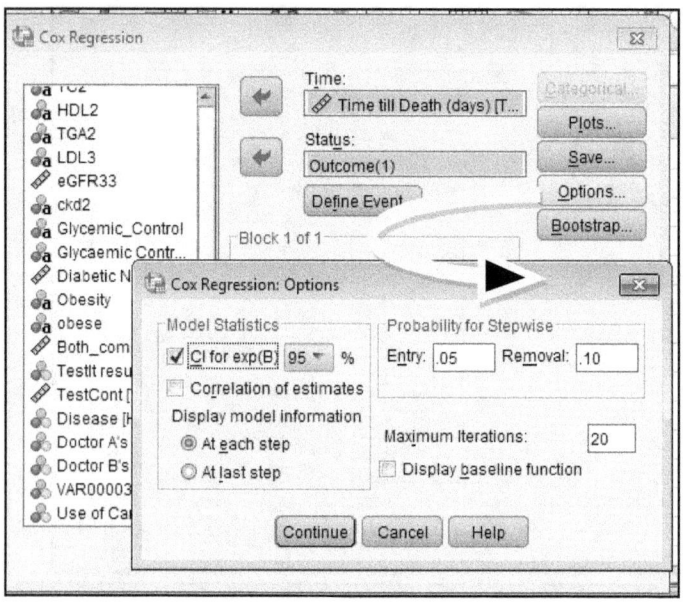

The results appear below:

Categorical Variable Codings[a,c]

		Frequency	(1)	(2)	(3)
MarStat[b]	0=Single	45	1	0	0
	1=Married	273	0	1	0
	2=Divorced	16	0	0	1
	3=Widowed	66	0	0	0
Education[b]	0=None	103	1	0	0
	1=Primary	108	0	1	0
	2=Secondary	112	0	0	1
	3=Tertiary	77	0	0	0

a. Category variable: MarStat (Marital Status)

b. Indicator Parameter Coding

c. Category variable: Education

Variables in the Equation

	B	SE	Wald	df	Sig.	Exp(B)	95.0% CI for Exp(B) Lower	Upper
MarStat			.793	3	.851			
MarStat(1)	.041	.271	.023	1	.879	1.042	.613	1.773
MarStat(2)	.028	.191	.022	1	.883	1.029	.708	1.495
MarStat(3)	-.261	.360	.524	1	.469	.770	.380	1.561
Education			10.517	3	.015			
Education(1)	.796	.260	9.355	1	.002	2.218	1.331	3.694
Education(2)	.473	.267	3.133	1	.077	1.605	.950	2.711
Education(3)	.473	.264	3.218	1	.073	1.604	.957	2.689
Cancidin	.041	.146	.081	1	.777	1.042	.783	1.388

The results show that use of Cancidin did not independently predict the risk of death. From the model, no education (coded as dummy variable "**Education(1)**") was the only predictor with a significant effect on outcome. These subjects with no education had an increased risk of death (Hazard Ratio: 2.22, 95% Confidence Intervals: 1.33 - 3.69).

BIBLIOGRAPHY

I found the following books and articles helpful:

- Peat J, Barton B. Medical Statistics: A Guide to Data Analysis and Critical Appraisal. 1st ed. Oxford: Blackwell Publishing Ltd; 2005.
- Burroughs TE. Assessing the Medical Literature. In: Schmitz P, Martin KJ, Miller DD, eds. Internal Medicine: Just the Facts: McGraw-Hill; 2008:73 - 89.
- Pallant J. SPSS SURVIVAL MANUAL: A step by step guide to data analysis using SPSS. 4th ed. Crows Nest: Allen & Unwin; 2011.
- Receiver Operating Characteristic. [25th July, 2015]; Available from: https://en.m.wikipedia.org/wiki/Receiver_operating_characteristic.
- Tyrell S. SPSS: Stats Practically Short and Simple: Ventus Publishing ApS; 2009

I also recommend the massive open online courses on data analysis available on www.alison.com, www.edx.org, and www.coursera.org.

FINAL WORDS...(FOR NOW)

I hope you have enjoyed reading (and using) this book as much as I enjoyed writing it.

There's a lot more to statistics and even the author is still learning!

I would love to hear your comments and criticism. Reach me on:

Twitter: @MedicAnalyst

Facebook: www.facebook.com/akemokwe.fatai

Email: drfatai998@gmail.com

Website: www.fatenigma.com

Cheers!

ABOUT THE AUTHOR

Dr Fatai Akemokwe (Diploma in Statistics, MBBS, MWACP) is a physician and clinical researcher . He is also a part-time instructor with Enigma Ventures, an organization focused on medical and information technology education. He considers himself a dabbling doctor with interests ranging from alternative rock music to philately.
He lives in Benin City, Nigeria.

FATAI AKEMOKWE